Journeyman Electrician Exam Prep.

If you are already confident and familiar with general electrical principles and terms, feel free to jump into section two for the practice exam questions on Page 59

Introduction:

The purpose of this guide is to equip you with the knowledge and skills you need to pass the journeyman electrician exam. As a student of electrical engineering, you already have a basic understanding of electrical concepts and theories. However, the journeyman electrician exam is a comprehensive test that covers a wide range of topics, from electrical services and wiring methods to special occupancies and renewable energy technologies. To succeed on the exam, you need to have a deep understanding of these topics and the ability to apply what you have learned in real-world situations.

The guide is designed to help you do just that. It starts with the basics of electrical formulas, such as Ohm's Law, and the relationship between power, current, and power factor. From there, it moves on to more advanced topics like conductor sizing and protection, motors, transformers, voltage drop, and over-current protection. The guide also covers the NEC codes and formulas you will need to know for the exam and practical calculations for residential and commercial loads.

By breaking down these complex topics into simple, easy-to-understand concepts, this guide will help you gain a deeper understanding of the material. Additionally, clearly, and concisely presenting information will help you remember what you have learned and apply it when it matters most – on exam day. So, use this guide as your foundation for success on the journeyman electrician exam. Remember that with hard work and dedication, you can achieve your goal of becoming a certified electrician.

The journeyman electrician exam is a critical step for anyone seeking to work as a licensed electrician. This exam is designed to test your knowledge of electrical theory, electrical codes, and electrical safety practices. The exam typically consists of multiple-choice questions that assess your understanding of topics such as electrical services, wiring methods, electrical equipment, motors and generators, and electrical control devices. To prepare for the exam, you'll need to have a solid understanding of electrical formulas, calculations, and NEC codes, as well as the ability to interpret electrical plans and blueprints. Additionally, it's important to have a thorough understanding of the electrical principles and practices that are covered in the National Electrical Code (NEC), as this codebook is the standard for electrical installations and practices in the United States. However, don't be intimidated! With the right study materials and preparation, you can feel confident and well-equipped to take the journeyman electrician exam.

As you begin this journey, I would like to emphasize the importance of proper preparation for your exam. While it may seem like a straightforward test of your technical knowledge, the reality is that the way you present yourself and approach the exam can greatly impact your performance. Firstly, it's crucial to study thoroughly and familiarize yourself with the material covered in the exam. This means memorizing facts and understanding the underlying concepts and theories. Secondly, it's important to practice with sample exams and review the actual test format so you can be confident and prepared on the day of the exam. Additionally, a calm and professional demeanor throughout the exam will show the evaluators that you take your future as a journeyman electrician seriously. Remember, the exam is an opportunity to showcase your skills and knowledge, so take it seriously and put your best foot forward!

Chapter 1: Electrical Fundamentals and theory.

As a Journeyman Electrician, it is important to have a solid understanding of various electrical concepts, definitions, formulas, and calculations. Here are some of the most important ones you should know:

Definitions:

- Voltage (V): The electric potential difference between two points, measured in volts (V)
- Current (I): The flow of electric charge, measured in amperes (A)
- Resistance (R): The opposition to the flow of electric current, measured in ohms (Ω)
- Power (P): The rate at which energy is consumed or converted, measured in watts (W)
- Capacitance (C): The ability of a component to store electric charge, measured in farads (F)
- Inductance (L): The property of an electrical component to store energy in a magnetic field, measured in henrys (H)

Formulas:

- Ohm's Law: $V = IR$ (Voltage equals current multiplied by resistance)
- Power: $P = IV$ (Power equals current multiplied by voltage)
- Energy: $E = VIt$ (Energy equals voltage multiplied by current multiplied by time)
- Capacitance: $C = Q/V$ (Capacitance equals charge divided by voltage)
- Inductance: $L = \Phi/I$ (Inductance equals magnetic flux divided by current)

Calculations:

- Series Circuits: Resistance in a series circuit is equal to the sum of all resistors in the circuit
- Parallel Circuits: Resistance in a parallel circuit is equal to the reciprocal of the sum of the reciprocals of each resistor in the circuit
- Series-Parallel Circuits: A combination of both series and parallel circuits, where the resistance calculation is more complex and requires a combination of the two formulas mentioned above.

It's important to familiarize yourself with these definitions, formulas, and calculations, as well as understand how to apply them in real-world scenarios.

As your "journeyman exam professor," I'm so thrilled to share with you the fundamentals and theory of electricity! Now, before we dive in, let me explain that electricity is the flow of electrons and is what powers our homes, our gadgets, and even our bodies. So, it's pretty important stuff!

Let's start with the basics. Electricity is the flow of electric charge from one point to another. It's a form of energy that is produced by the movement of charged particles, such as electrons. This movement of electrons is what generates the flow of electrical current.

Now, we can't see or touch electricity, but we can certainly feel its effects! That's because electricity produces heat, light, and magnetic fields, among other things. And that's why we use it to power so many different things in our lives.

Next, let's talk about electrical potential difference, also known as voltage. Voltage is the measure of the electric potential energy per unit charge. It's what drives the flow of electrons through a conductor, like a wire. And the more voltage you have, the more energy you have available to do work.

Now, when we talk about electrical circuits, we need to think about the concept of resistance. Resistance is the measure of the opposition to the flow of current in a circuit. This opposition to flow is what generates heat, which can damage electrical components and cause electrical fires.

So, that's where Ohm's law comes in. It states that the current flowing through a conductor between two points is directly proportional to the voltage across the two points, and inversely proportional to the resistance between them. This is a fundamental relationship that is widely used in electrical engineering to design and analyze electrical circuits.

Finally, let's talk about alternating current (AC) and direct current (DC). AC is an electrical current that changes direction periodically, while DC is an electrical current that flows in only one direction. Most of the electrical power that we use in our homes is AC because it can be transmitted over long distances with less loss of energy than DC.

So, there you have it! These are just the basics of electrical fundamentals and theory, but they form the foundation for so much more that you'll need to know as a journeyman electrician.

electricity is a fascinating and complex subject that is at the heart of our modern world. To start off with the basics, electricity is the flow of charged particles, typically electrons, through a conductor such as a metal wire. These charged particles are what make up the electrical current that we use to power all sorts of devices in our homes, workplaces, and beyond.

But where does this flow of charged particles come from? Well, that's where the concepts of voltage, resistance, and current come into play. Voltage, also known as electrical potential difference, is the force that pushes the charged particles through the conductor. Resistance, on the other hand, is the opposition to the flow of current in the conductor. And current, of course, is the actual flow of charged particles.

Now, it's important to note that these three concepts are closely related and interdependent. A change in one of them will affect the others. For example, increasing the voltage will increase the current flow, but it will also increase the resistance. It's like when you're trying to force water through a narrow pipe - the water will flow faster, but it will also face greater resistance.

Another important concept in electricity is electrical power, which is the rate at which energy is used or transferred. This is where Ohm's Law comes in, which states that the power in an electrical circuit is equal to the voltage multiplied by the current. This relationship is crucial for electricians to understand, as it helps them determine the size of the power source needed for a given circuit.

And then there's the matter of electrical circuits themselves. A circuit is a closed loop of conductors through which current can flow. Circuits can be simple or complex, and they can contain various components such as resistors, capacitors, inductors, and transistors. Understanding how these components work and interact with each other is crucial for electricians, as it allows them to design and troubleshoot electrical systems.

Now **voltage**, what a fascinating concept! As an electrician, it's essential that you understand this fundamental concept of electricity. Voltage, also known as electric potential difference, is the measure of electrical energy that drives the flow of electric current in a circuit.

Think of it like water pressure in a hose. The water pressure is what pushes the water through the hose, just as the voltage is what drives the electric current through a circuit. The greater the voltage, the greater the energy and the more current that can flow.

Now, here's where it gets a bit perplexing: voltage is not the same thing as current. Voltage is the measure of potential energy, while current is the measure of the actual flow of electrons.

These two concepts are often confused, but it's important to understand the difference in order to fully grasp the workings of an electrical circuit.

You'll also come across the term "electromotive force," or EMF, when discussing voltage. This term is often used interchangeably with voltage, but there is a subtle difference. EMF refers to the source of voltage, such as a battery or generator, while voltage refers to the actual measure of energy in the circuit.

And just when you thought you had a handle on voltage, there's one more important piece of information to remember: voltage is measured in volts (V). This is a standard unit of measurement that you'll come across frequently as you study electrical systems and take the journeyman electrician exam.

So, to summarize, voltage is a measure of the electrical energy that drives the flow of current in a circuit. It is not the same thing as current, but rather the measure of potential energy. Understanding voltage and its relationship to current is a crucial aspect of understanding electricity and passing the journeyman electrician exam.

When it comes to electrical concepts, "current" is one of the most fundamental and important to understand. So buckle up, because I'm about to dive into the nitty-gritty of this essential idea!

At its core, current refers to the flow of electric charge through a conductor. This flow is usually measured in units of Amperes, or "Amps" for short. Now, let's say you've got a light bulb connected to a battery. When you turn the light bulb on, electric charge is flowing from the battery, through the light bulb, and back to the battery. That flow of charge is the current, and it's what allows the light bulb to produce light and heat.

Now, let's talk about how resistance affects current. Resistance is a measure of how much a material resists the flow of electric charge. Different materials have different levels of resistance, and that resistance can change depending on a number of factors, such as temperature and the presence of other materials. When a material has high resistance, it resists the flow of electric charge and reduces the current.

To understand how this works in real-world applications, it's helpful to think about an electric circuit as a pipe filled with water. Imagine the water is electric charge and the pipe is a conductor. If you increase the diameter of the pipe, the water will flow more easily, just as increasing the cross-sectional area of a conductor will reduce its resistance and allow more current to flow. Conversely, if you reduce the diameter of the pipe, the water will flow more slowly, and the same goes for reducing the cross-sectional area of a conductor and increasing its resistance.

Now, when it comes to your journeyman electrician exam, it's important to have a solid understanding of Ohm's Law, which states that the current in a circuit is directly proportional to the voltage across the circuit and inversely proportional to the resistance of the circuit. In mathematical terms, this is expressed as:

$$I = V / R$$

Where I is the current in Amps, V is the voltage in Volts, and R is the resistance in Ohms. This equation is one of the most important in all of electrical engineering, and you'll definitely want to have it memorized for your exam.

So there you have it! In a nutshell, current is the flow of electric charge through a conductor, and its flow is affected by the resistance of the conductor. Understanding these concepts and having a solid grasp of Ohm's Law will be key to your success on your journeyman electrician exam.

Ah, **amps**, Let's start with the basics, shall we? An "amp" or "ampere" is the unit of measurement for electrical current. You see, electricity is like a river of electrons flowing through a conductor, like a wire. The number of electrons flowing through that wire per second is what we call the current, and we measure it in amps.

Here's where it gets a little perplexing: the amount of current flowing through a conductor is not constant but can change based on various factors such as the resistance of the conductor, the voltage applied, and even the temperature of the conductor.

Let's put it in simple terms: imagine a hose, the water flowing through the hose can be thought of as the electrical current. The more water you pump through the hose, the stronger the current. Similarly, if you increase the voltage, it's like increasing the water pressure, which will increase the amount of current flowing through the conductor.

It's important to understand amps, as they play a crucial role in determining the amount of power that can be delivered to a device. For example, a high-powered motor may require a large amount of current to operate, whereas a simple light bulb may only require a small amount of current.

So, to summarize, amps are the unit of measurement for electrical current, which can be affected by factors such as voltage and resistance and play a crucial role in determining the power that can be delivered to a device.

As a journeyman electrician preparing for your exam, it's important to understand the concept of **resistance**. Resistance is a measure of how much a material resists the flow of electrical current. This is an essential concept in the electrical field, as it has a direct impact on the safety and efficiency of electrical systems.

The unit of measurement for resistance is the ohm (Ω). It's defined as the amount of voltage required to drive one ampere of current through a material with a resistance of one ohm. To put it simply, the higher the resistance of a material, the more energy it takes to push the same amount of electrical current through it.

In practical terms, resistance plays a crucial role in electrical circuits. For example, in a circuit with a high resistance, less current flows, and the circuit becomes less efficient. On the other hand, a circuit with a low resistance allows more current to flow and is more efficient. However, a circuit with too low of a resistance can also be dangerous, as it can cause a large amount of electrical current to flow, which can lead to fires or other hazards.

When it comes to preparing for your journeyman electrician exam, it's essential to understand the codes and standards that relate to resistance. For example, the National Electrical Code (NEC) specifies minimum resistance values for electrical conductors to ensure that they can safely carry the electrical current without overheating. It's also important to understand the real-world applications of resistance and how it affects the electrical systems you'll be working on as a journeyman electrician.

In summary, resistance is a fundamental concept in the electrical field, and a deep understanding of it is crucial for preparing for your journeyman electrician exam. Make sure to study the codes and standards related to resistance, as well as real-world applications, to give yourself the best chance of success on exam day.

Are amps and current the same thing?

No, amps and current are not exactly the same thing.

Current is defined as the flow of electric charge in a circuit. The unit of measurement for current is the Ampere, often referred to as an "Amp" for short. So, when you measure current, you are measuring the flow of electric charge in Amps.

So in a way, you can say that Amps are the units used to measure the current flow in a circuit. It's important to understand the relationship between Amps, Current, and other electrical quantities like Voltage and Resistance, as they all work together to define the behavior of an electrical circuit.

How amps and volts are related but different.

Let's start with a definition of volts and amps. Volts, or voltage, is a measure of electrical potential difference between two points in a circuit. It is the force that drives the flow of electric current, kind of like the water pressure that pushes water through a hose. On the other hand, amps, or amperage, is a measure of the flow rate of electric current in a circuit, or the amount of electric charge that moves through a circuit per second. It's like the amount of water that flows through a hose in a given amount of time.

Now, let's see how these two concepts are related but different. To put it simply, voltage is like the water pressure and amps are like the flow rate. Just like how you need enough water pressure to get water to flow through a hose, you also need enough voltage to get current to flow in a circuit. But just like how you can have high water pressure but low flow rate, you can also have high voltage but low amperage. Similarly, you can have low water pressure but high flow rate, or low voltage but high amperage.

In practical terms, this means that a high voltage source may not necessarily have a lot of current flowing through it, and vice versa. Understanding this relationship is important in electrical work, because it affects the amount of power that can be transmitted, the amount of energy that can be stored, and the amount of heat that is generated in a circuit.

It is also important to note that when we talk about electrical power, it is the product of volts and amps, or P=V*I. So, even if the voltage is high, but the amperage is low, the total power being transmitted may not be high. On the other hand, if the voltage is low but the amperage is high, the total power being transmitted may still be high. This is why it's important to understand both voltage and amperage when working with electrical systems.

Ah, the **circuit**! This is one of the most fundamental concepts you'll need to understand for your journeyman electrician exam. In simple terms, a circuit is a closed path which electrical current flows through. It's like a loop - electricity flows in one direction and then back again, over and over.

Now, there are many different types of circuits you'll encounter in your electrical work. There are series circuits, where components are connected end-to-end in a single path, and parallel circuits, where components are connected to the same voltage source, but split off into separate branches. There are also DC circuits, which use direct current, and AC circuits, which use alternating current.

One important factor to consider when working with circuits is resistance. Resistance is the measure of how difficult it is for electrical current to flow through a material. Every material has some level of resistance, and you'll need to be able to calculate and work with resistance values to design and troubleshoot electrical circuits.

It's also crucial to understand Ohm's Law, which states that the current through a conductor between two points is directly proportional to the voltage across the two points, and inversely proportional to the resistance between them. This law is often expressed as the formula V=IR, where V is voltage, I is current, and R is resistance.

When preparing for your journeyman electrician exam, it's important to become familiar with the National Electric Code (NEC), which lays out specific guidelines and requirements for electrical circuits and installations. There are many specific rules and regulations you'll need to be familiar with, such as those relating to circuit sizing, overcurrent protection, grounding, and more.

In terms of real-world examples, consider a simple circuit like a light bulb in your home. The bulb is connected to a power source, such as a wall outlet, and electricity flows through the bulb's filament, creating light. The resistance of the filament affects the amount of current that can flow through the circuit, and the amount of light produced by the bulb.

In conclusion, circuits are essential to the electrical trade, and understanding the principles of resistance, Ohm's Law, and code requirements will be key to your success on the journeyman electrician exam. So don't be afraid to dive into these concepts and really get to know them - with time and practice, you'll be able to tackle any circuit problem that comes your way!

Capacitance is an important concept in electrical engineering and is essential for a journeyman electrician to understand. In simple terms, capacitance is the ability of a device to store electrical energy in an electric field. This stored energy can then be released back into the circuit as needed. The amount of capacitance in a circuit is measured in Farads, and it is a property of the components used in the circuit.

When preparing for the journeyman electrician exam, it is important to understand how capacitance is related to other electrical parameters, such as voltage and current. The relationship between these parameters can be described by Ohm's Law and the Capacitor equation, which states that the voltage across a capacitor is proportional to the charge stored on it. This relationship is crucial in the design of electrical systems and in the diagnosis of electrical problems.

In practical terms, capacitors are used in many electrical applications, including power factor correction, energy storage, and signal filtering. Understanding how capacitors work and how they can be used in circuits is a key part of preparing for the journeyman electrician exam. You may also be asked questions about specific code references, such as the National Electric Code (NEC), which provides guidelines for the safe installation and use of electrical systems.

So, it is important to study and understand capacitance, its properties and its applications, as it will help you in solving electrical problems and will also come handy in your journeyman electrician exam. To sum up, capacitance is a fundamental concept in electrical engineering and is an essential part of your preparation for the journeyman electrician exam.

In electricity**, resistance** refers to the opposition that a material offers to the flow of electric current. It's measured in Ohms (Ω), and the formula for calculating resistance is: R = V / I, where R is resistance, V is voltage, and I is current.

One of the most important aspects of resistance is that it can cause energy to be lost in the form of heat, which is why electrical components such as resistors are often used to regulate the flow of current in circuits.

When preparing for your journeyman electrician exam, it's important to be familiar with the National Electrical Code (NEC), which provides safety standards for electrical installations. One of the NEC requirements is that all electrical equipment and systems must be installed in such a way as to minimize electrical hazards and fire risks. This includes ensuring that electrical circuits are properly protected against overloading, which can occur when too much current flows through a circuit with a limited resistance.

In real-world applications, resistance can affect the performance of electrical systems, such as motors, transformers, and generators. For example, the resistance of the wire used in a circuit affects the amount of current that can be safely carried by the wire without overheating.

So, it's important to understand the concept of resistance, as well as how it relates to voltage, current, and the NEC safety requirements for electrical installations. This knowledge will help you successfully pass your journeyman electrician exam and work safely and effectively as an electrician.

Understanding electrical **power** is an essential part of preparing for your journeyman electrician exam. Power can be defined as the rate at which energy is being used or generated. In electrical terms, it is the rate at which electrical energy is being transferred from one place to another or being transformed from one form to another.

The standard unit of electrical power is the watt (W). Power can also be expressed in kilowatts (kW), which is equal to 1000 watts, or in megawatts (MW), which is equal to one million watts.

There are two important formulas related to electrical power that you should know:

Power (P) = Voltage (V) x Current (I): This formula shows the relationship between voltage, current, and power in a circuit. The formula states that the power in a circuit is equal to the voltage across it multiplied by the current flowing through it.
Power (P) = Current (I)^2 x Resistance (R): This formula is known as Ohm's law, and it states that the power in a circuit is equal to the square of the current flowing through it multiplied by the resistance of the circuit.
It's important to understand the difference between these two formulas and when to use each of them. The first formula is used when the voltage and current in a circuit are known, and the power needs to be calculated. The second formula is used when the current and resistance in a circuit are known, and the power needs to be calculated.

In real-world applications, it's essential to understand electrical power because it helps electricians to determine the size and type of electrical equipment that is needed for a specific job. For example, if you need to power a large motor, you will need to know the power requirements of the motor so that you can select the appropriate electrical equipment.

inductance is a property of an electrical circuit that opposes changes in current. It's measured in henries (H). Inductance occurs when a change in current flow through a coil of wire creates a magnetic field. The magnetic field induces an electromotive force (EMF) that opposes the change in current.

In a circuit, inductance can be thought of as a type of "inertia" for electricity. It resists changes in the flow of current and tends to keep the current flowing in the same direction. This property is particularly important in AC circuits, where the direction of current changes constantly.
When preparing for your journeyman electrician exam, it's important to understand the basic properties of inductance, including:

- Self-inductance: This occurs when an inductor induces a voltage in itself due to a change in current flow.
- Mutual inductance: This occurs when two or more inductors are near each other and their magnetic fields interact, inducing a voltage in each other.
- Inductive reactance: This is the opposition to current flow that occurs in an inductive circuit. It's equal to $2\pi fL$, where f is the frequency of the AC current and L is the inductance.
- Inductive coupling: This occurs when energy is transferred from one coil of wire to another through the magnetic field they produce.

In real-world applications, inductors are often used in electrical circuits to store energy in the form of a magnetic field and then release it as needed. They're also used to filter and regulate the flow of electrical energy.
Some examples of the use of inductors include:

- Chokes: These are inductors used to block high-frequency signals while allowing low-frequency signals to pass.
- Transformers: These are devices that use inductors to transfer energy from one circuit to another.
- Inductive sensors: These are devices that use the magnetic field produced by an inductor to sense the presence of an object.

Ohm's law is one of the most fundamental principles in electricity, and it's essential that you understand it well if you're preparing for your journeyman electrician exam. Ohm's law states that the current flowing through a conductor between two points is directly proportional to the voltage across the two points and inversely proportional to the resistance between them. The equation for Ohm's law is given by:

$$V = IR$$

where V is the voltage in volts, I is the current in amps, and R is the resistance in ohms.

It's important to remember that Ohm's law is only applicable to linear, passive electrical circuits. In these types of circuits, the relationship between voltage, current, and resistance is constant, and the current will change in direct proportion to the voltage if the resistance remains constant.

To put this into a real-world example, consider a 100-watt light bulb with a resistance of 120 ohms. If you know the voltage of the circuit is 120 volts, you can use Ohm's law to calculate the current flowing through the bulb:

$$I = V / R = 120 / 120 = 1 A$$

This means that there is 1 ampere of current flowing through the light bulb when it's turned on.

It's also worth noting that the power of a circuit can be calculated using Ohm's law as well. The power is given by:

$$P = VI = I^2 R = V^2 / R$$

Where P is the power in watts, I is the current in amps, V is the voltage in volts, and R is the resistance in ohms.

In conclusion, understanding Ohm's law and its relationship to voltage, current, and resistance is crucial for success on the journeyman electrician exam. Make sure you're comfortable with the formula and the concept before taking the test!

Energy is an essential concept in electrical engineering and is fundamental to the understanding of how electrical systems work. It refers to the ability to do work, and in electrical systems, it's typically measured in units of Joules (J).

The relationship between energy, power, and time is described by the equation: Energy = Power * Time. This equation is important for understanding how electrical energy is consumed in a system, and how much energy is required to perform a certain task.

When preparing for your journeyman electrician exam, it's important to understand the different forms of energy, including electrical energy, thermal energy, and mechanical energy. Electrical energy is the energy stored in an electric field or in an electric circuit, and it's directly related to the voltage and current in the circuit. Thermal energy is the energy associated with the temperature of a material, and mechanical energy is the energy stored in an object due to its motion or position.

It's also important to understand how energy is transformed in electrical systems. For example, electrical energy can be transformed into heat, light, or mechanical energy. The efficiency of energy conversion in electrical systems is expressed as a percentage, and it's a measure of how much of the electrical energy is transformed into useful energy, rather than being lost as heat or other forms of waste energy.

When preparing for your journeyman electrician exam, it's also important to be familiar with the electrical safety codes and standards, such as the National Electrical Code (NEC), that govern the use and installation of electrical systems. These codes and standards are designed to ensure the safe and efficient use of electrical energy, and they include requirements for the installation of electrical systems, equipment, and devices.

Overall, understanding the principles of energy is critical for success on the journeyman electrician exam, and it's essential for ensuring the safe and efficient use of electrical energy in the real world.

Kirchhoff's laws are fundamental principles that describe the behavior of electrical circuits. There are two laws, known as Kirchhoff's voltage law (KVL) and Kirchhoff's current law (KCL). Kirchhoff's voltage law (KVL) states that the total voltage around any closed loop in a circuit is equal to zero. In other words, the sum of all the voltage drops in a loop must equal the sum of all the voltage gains. This law is useful in analyzing the behavior of complex circuits by breaking them down into smaller, simpler parts.

Kirchhoff's current law (KCL) states that the total current entering a node (a junction of two or more wires) in a circuit must equal the total current leaving that node. In other words, the sum of all the currents flowing into a node must equal the sum of all the currents flowing out of the node.

Kirchhoff's Laws: There are two laws, known as Kirchhoff's Current Law (KCL) and Kirchhoff's Voltage Law (KVL).

KCL states that the sum of all currents entering a junction (a node) must equal the sum of all currents leaving the node. In other words, the total current flowing into a node is equal to the total current flowing out of it. Mathematically, this is expressed as the sum of all currents entering the node must equal zero: $\sum I\_in = 0$.

KVL states that the sum of the voltages around a closed loop must equal zero. In other words, the total voltage gain around a loop must equal the total voltage loss. Mathematically, this is expressed as the sum of voltages around a loop must equal zero: $\sum V\_loop = 0$.

Knowing Kirchhoff's laws is critical for your journeyman electrician exam because they form the basis of many electrical calculations and circuit analysis techniques. These laws are used to calculate voltage drops, current flows, and power in a circuit, and they are essential for understanding the behavior of AC and DC circuits.

In the National Electric Code (NEC), Kirchhoff's laws are not explicitly referenced, but the concepts behind these laws are used to ensure the safe design and operation of electrical systems. The NEC sets guidelines for calculating wire size, breaker sizing, and other electrical calculations that are based on Kirchhoff's laws.

Here are a few examples to help illustrate the concepts behind Kirchhoff's laws:

1. Example 1: Consider a simple circuit with a battery and a light bulb. According to KVL, the sum of the voltage drops around the circuit must equal the voltage of the battery.
Voltage drop across the light bulb + voltage drop across the wire = Voltage of the battery
2. Example 2: Consider a node with three wires connected to it. According to KCL, the sum of the currents flowing into the node must equal the sum of the currents flowing out of the node.
Current flowing into the node 1 + current flowing into the node 2 + current flowing into the node 3 = Current flowing out of the node 1 + current flowing out of the node 2 + current flowing out of the node 3

impedance is a measure of the opposition that an electrical circuit presents to the flow of an alternating current (AC). It's a complex quantity that includes both resistance and reactance (the opposition to changes in current due to inductance and capacitance in the circuit). In other words, impedance is a combination of resistance and reactance that represents the total opposition to AC flow in a circuit.

Impedance can be expressed in ohms and is denoted by the symbol Z. The formula for impedance is:

$$Z = \sqrt{R^2 + X^2}$$

Where R is resistance and X is reactance.

It's important to understand that impedance is not the same thing as resistance. Resistance only opposes the flow of direct current (DC) in a circuit. However, impedance opposes the flow of AC in a circuit and takes into account both resistance and reactance.

Knowing impedance is important for journeyman electricians because it helps you understand the behavior of AC circuits. For example, if you know the impedance of a circuit, you can calculate the current, voltage, and power that are present in the circuit. You can also use impedance calculations to determine the maximum power that can be transferred from one point in a circuit to another.

So, to sum up, impedance is an important concept to understand for your journeyman electrician exam. It's the total opposition to AC flow in a circuit and includes both resistance and reactance. To calculate impedance, you can use the formula $Z = \sqrt{R^2 + X^2}$, where R is resistance and X is reactance.

The individual who helped me prepare for the journeyman electrician exam was big on helping me understand the basics through analogies that made me say, "ohhhhhh.. now I get it". Here are some of those basic examples.

1. **Ohm's Law: Ohm's Law states that the current flowing through a conductor between two points is directly proportional to the voltage across the two points and inversely proportional to the resistance between them. Analogy: Imagine a water hose with a nozzle attached to it. The water pressure in the hose is like the voltage, the amount of water flowing through the nozzle is like the current, and the nozzle's opening size is**

like the resistance. If you increase the pressure (voltage), more water (current) will flow through the nozzle (resistance).

2. Resistance: Resistance is a measure of how much a material resists the flow of electrical current. Analogy: Imagine a swimming pool filled with molasses. The molasses represents resistance because it slows down the flow of water (electrical current).

3. Current: Current is the flow of electrical charge in a circuit. Analogy: Imagine a river with water flowing through it. The flow of water in the river is like the flow of electrical charge in a circuit.

4. Voltage: Voltage is the electrical potential difference between two points. Analogy: Think of voltage as a hill. The higher the hill, the more potential energy an object has. Similarly, the greater the voltage difference between two points, the greater the electrical potential difference.

5. Power: Power is the rate at which electrical energy is converted into another form of energy. Analogy: Power can be thought of as the speed at which a machine is working. If a machine is working faster, it has more power, just like an electrical circuit that is converting electrical energy into another form of energy more quickly has more power.

6. Capacitance: Capacitance is the ability of a material to store electrical energy. Analogy: Think of capacitance as a battery. Just as a battery can store energy, a material with capacitance can store electrical energy.

7. Inductance: Inductance is the ability of a coil of wire to store energy in a magnetic field. Analogy: Think of inductance as a spring. Just as a spring can store energy, a coil of wire with inductance can store energy in a magnetic field.

8. Kirchhoff's Laws: Kirchhoff's Laws describe how electrical energy is conserved in a circuit. Analogy: Think of Kirchhoff's Laws as a game of catch with a ball. The total amount of energy in the ball (electrical energy in the circuit) must be conserved, meaning that the energy in the ball cannot be created or destroyed, it can only be transferred from one person to another.

9. Impedance: Impedance is the total opposition to the flow of electrical current in an AC circuit. Analogy: Think of impedance as a speed bump. Just as a speed bump slows down the speed of a car, impedance slows down the flow of electrical current in an AC circuit.

CHAPTER 2: Electrical Services, Equipment, and Systems.

As a journeyman electrician, it's important to have a solid understanding of **electrical services, equipment, and systems**. This includes everything from the main electrical service that brings power into a building, to the various pieces of equipment that make up the electrical distribution system, to the various electrical systems that are installed in a building, such as lighting, fire alarms, and HVAC systems.

Let's start with the main electrical service. This is typically made up of a service entrance conductor, which brings power into the building, and a service disconnecting means, which is used to disconnect the power in the event of an emergency or maintenance. The size of the service entrance conductor is determined by the amount of power that's needed for the building, and is typically specified in amps.

Next, we have the electrical distribution system, which is made up of various pieces of equipment, such as circuit breakers, panels, transformers, and busways. Circuit breakers are used to protect the electrical system from overloading, and come in different sizes, with the most common being the 20-amp, 30-amp, and 40-amp breakers. Panels, also known as load centers, house the circuit breakers and provide a centralized location for the distribution of power.

Transformers are used to step down the voltage of the electrical service so that it can be used safely within the building. They come in different sizes, with the most common being the single-phase and three-phase transformers. Busways are used to distribute power within a building and are typically specified in amps.

Finally, we have the various electrical systems that are installed in a building, such as lighting, fire alarms, and HVAC systems. Lighting systems are used to provide illumination and come in different types, such as incandescent, fluorescent, and LED lighting. Fire alarm systems are used to alert people in the event of a fire and come in different types, such as conventional and addressable fire alarm systems. HVAC systems are used to control the temperature and air quality within a building and come in different types, such as split systems and packaged systems.

As you prepare for your journeyman electrician exam, it's important to familiarize yourself with these electrical services, equipment, and systems and to know the codes and regulations that apply to each. Knowing the NEC (National Electrical Code) is also crucial in order to ensure that electrical installations are safe and meet the minimum standards set forth by the code.

The main electrical service is the main source of electrical power in a building. It is the main point of entry for electrical energy into the building, and it's where the electrical power is distributed to the various electrical circuits and systems within the building. Understanding the main electrical service is crucial for electricians, especially those who are preparing for their journeyman electrician exam. Here are some key points you should know about the main electrical service:

1. Service Entrance Cable: This is the cable that brings electrical power into the building from the utility company. It can be either underground or overhead.
2. Service Entrance Panel: This is the main electrical panel that is located near the service entrance cable. It contains the main circuit breaker or fuse that protects the entire building from overloading.
3. Main Circuit Breaker or Fuse: This device is used to protect the entire building from overloading. It's the first line of defense against electrical fires.
4. Service Entrance Rated Capacity: The main electrical panel should be rated for the amount of electrical power that the building requires. This is called the service entrance rated capacity and is usually measured in amperes (amps).
5. Grounding System: The main electrical service also includes a grounding system. This is a safety system that provides a path for electrical current to flow in the event of a fault or short circuit.
6. Neutral and Ground Wiring: The neutral and ground wiring are two separate conductors that are used in the main electrical service. The neutral conductor carries the electrical current back to the source while the ground conductor provides a path for electrical current to flow in the event of a fault.
7. Overcurrent Protection: The main electrical service must have overcurrent protection to prevent electrical fires caused by overloading. This can be provided by either fuses or circuit breakers.

Now, let's talk about some real-world examples. If you have ever been to a construction site, you may have seen a large electrical panel that is used as the main electrical service for the building. This panel will contain the main circuit breaker or fuse and will be rated for the amount of electrical power that the building requires.

Another example is in a residential home. The main electrical service panel in a residential home is typically located in a basement or garage. It contains the main circuit breaker or fuse that protects the entire home from overloading. The panel is usually rated for 100 amps or 200 amps, depending on the size of the home.

Let me give you a deep dive into **electrical distribution systems**, which are crucial for delivering electrical power to the end-users in a safe and efficient manner.

A typical electrical distribution system consists of several components, including the main service, transformers, switchgear, overcurrent protection devices, and distribution panels. The main electrical service is responsible for bringing electrical power from the utility company's power grid into a building or facility. The transformers then reduce the voltage to a level that can be used safely for electrical distribution within the building.

The switchgear is a collection of switchboards, circuit breakers, and fuses that control and protect the electrical power as it flows through the distribution system. Overcurrent protection devices, such as fuses and circuit breakers, protect the system against short circuits, ground faults, and overloading.

Distribution panels, also known as load centers or breaker panels, distribute the electrical power to different circuits within a building. These panels typically contain multiple circuit breakers that allow individual circuits to be controlled and protected.

It is important for electricians to understand the different components and their functions within the electrical distribution system, as well as the electrical codes and standards that govern the design and installation of these systems. The National Electric Code (NEC) is a commonly referenced set of standards for electrical installations, and it provides guidelines for the design and installation of electrical systems.

Some of the specific NEC codes you should be familiar with for your journeyman electrician exam include:

- NEC 310-16, which sets standards for wire sizing based on ampacity and the conductor's insulation type.
- NEC 408-3, which provides guidelines for the sizing and installation of overcurrent protection devices.
- NEC 240-3, which establishes minimum standards for overcurrent protection.
- NEC 210-20, which requires grounding and bonding of electrical systems.

You should also be familiar with the electrical distribution system's design and installation requirements, as well as the maintenance and repair procedures for electrical distribution systems.

In real-world applications, electricians may be involved in the design, installation, maintenance, and repair of electrical distribution systems in commercial, industrial, and residential settings. For example, an electrician might install a new electrical distribution system in a new building, upgrade an existing system in an older building, or repair a damaged system in a facility.

In conclusion, understanding electrical distribution systems and their components is a crucial aspect of being a successful electrician, and it is essential for passing your journeyman electrician exam. Make sure to study and understand the NEC codes related to electrical distribution systems, and be familiar with real-world examples of electrical distribution system design, installation, maintenance, and repair.

Let's dive into the different **electrical systems in buildings**.

Starting with Lighting Systems:
Lighting systems play a crucial role in building electrical systems, as it provides not only the visibility required for daily activities but also contributes to the ambiance of a space. You need to be familiar with the different types of lighting fixtures, such as incandescent, fluorescent, LED, and their characteristics. Also, you need to understand how to properly install lighting systems in compliance with the National Electrical Code (NEC) and local building codes. When it comes to the NEC, you should be familiar with the requirements for overcurrent protection, lighting outlet locations, and wire sizing for lighting circuits.

Moving on to Fire Alarm Systems:
Fire alarm systems are essential for ensuring the safety of the building's occupants in the event of a fire. You need to be familiar with the different types of fire alarms, including conventional, addressable, and analogue, and the specific requirements for their installation. The NEC provides the minimum requirements for fire alarms, including those for power supplies, backup power, and smoke detectors.

Finally, HVAC Systems:
Heating, Ventilation, and Air Conditioning (HVAC) systems are critical for maintaining a comfortable indoor environment. As a journeyman electrician, you need to be familiar with the different types of HVAC systems, including central air systems, ductless mini-split systems, and

heating-only systems. Additionally, you should understand the electrical requirements for HVAC systems, including power supply, control circuits, and overcurrent protection. You should also be familiar with the NEC requirements for HVAC systems, including those for disconnecting means, conductor sizing, and grounding.

That should give you a good overview of the various electrical systems in buildings and what you need to know for your journeyman electrician exam. It's important to study these systems in depth and understand the details of the NEC requirements, as well as the practical considerations involved in their installation and maintenance.

A Separately Derived System is an electrical system that has no direct electrical connection to the power source of the primary supply, but instead gets its electrical power from a secondary or separate source. Examples of such secondary sources include generators, transformers, and battery systems. These systems provide electrical power to a single location, or multiple locations, and the voltage is derived from a single source, such as a generator or transformer.

It's important to know that a Separately Derived System must have its own neutral conductor, and its own grounding system, which is independent of the primary supply grounding system. The grounding system must be of sufficient size and properly installed to ensure safe operation of the electrical system.

Here are a few things you should know about Separately Derived Systems for your exam:

1. A Separately Derived System must have its own overcurrent protection.
2. The grounding system of a Separately Derived System must be bonded to the grounding system of the building.
3. The equipment grounding conductor of a Separately Derived System must be connected to the grounding electrode.
4. The neutral conductor of a Separately Derived System must be grounded.

In real-world examples, Separately Derived Systems are commonly used in emergency backup power systems, computer rooms, and data centers, where a constant and reliable power source is crucial. These systems are also used in industrial applications, where a separate power source is required for specific equipment or processes.

electrical feeders! They're a crucial component of any electrical distribution system. To prepare for your Journeyman electrician exam, it's important to understand what feeders are, how they work, and what you need to know about them. Let's dive in!

Electrical feeders refer to the electrical cables or conductors that provide power to an electrical distribution system. They play a vital role in transmitting electrical energy from the source of power to different loads within a building or facility. Feeders come in different sizes and types, and the choice of which to use depends on the power requirements of the loads they're supplying.

Here are a few key things to keep in mind when it comes to electrical feeders:

1. Sizing: Feeder sizing is important, as it ensures that the feeder can carry the maximum amount of current required by the loads it supplies without overheating or causing other electrical problems.
2. Voltage drop: Voltage drop is the decrease in voltage that occurs along a feeder due to resistance, inductance, and capacitance. It's important to consider voltage drop when selecting feeders to ensure that the loads receive the correct voltage level.
3. Protective devices: Feeders should be protected by overcurrent protective devices, such as circuit breakers or fuses, to prevent overloading and other electrical hazards.
4. Grounding: Feeders must be grounded properly to ensure electrical safety and to reduce the risk of electrical shock or fire.

When it comes to your Journeyman electrician exam, you'll likely be tested on your knowledge of electrical feeders and their applications, including the proper selection and sizing of feeders, voltage drop calculations, and grounding requirements. You should also be familiar with National Electrical Code (NEC) requirements for feeders, as well as any local or state codes that may apply.

1. NEC 310-16: This section provides the ampacity and size requirements for feeder conductors.
2. NEC 310-17: This section provides the ampacity and size requirements for feeder conductors supplying separate buildings or structures.
3. NEC 310-19: This section provides the requirements for feeder conductors that are supplied by separate services.
4. NEC 310-60: This section provides the requirements for feeder conductors for multi-wire branch circuits.

5. NEC 310-63: This section provides the requirements for feeder conductors for multi-wire branch circuits where overcurrent protection is located at the first overcurrent device.
6. NEC 215-2: This section provides the requirements for feeder conductor protection, such as overcurrent protection, grounding, and bonding.
7. NEC 230-79: This section provides the requirements for feeder conductor sizing and protection for feeding panelboards.

In addition to these sections, it's also a good idea to review NEC articles 310, 215, and 230 in their entirety to gain a comprehensive understanding of feeder requirements.

Real-world examples of feeders include the electrical cables that supply power to homes and buildings, as well as the electrical conductors that supply power to large industrial facilities. The size and type of feeders used in these applications will vary depending on the power requirements of the loads they supply.

Feeder sizing is a critical aspect of electrical design and installation. This involves determining the appropriate size of the conductors and circuit protective devices used in a feeder circuit based on the electrical load it is intended to serve.

Here are some key things you need to know about feeder sizing for your journeyman electrician exam:
1. Understanding of the NEC load calculation requirements: You need to have a good understanding of the National Electrical Code (NEC) load calculation requirements and how they relate to feeder sizing. You'll need to know the requirements for feeder conductors and overcurrent protective devices.
2. Knowledge of ampacity and conductor sizing: The ampacity of the feeder conductor is the maximum amount of current that can safely flow through the conductor without causing damage. You need to be familiar with the ampacity of different sizes of conductors and how to size conductors for a given load.
3. Awareness of the effects of ambient temperature: The ambient temperature surrounding the feeder conductor can affect its ampacity. You need to be aware of the effect of ambient temperature on feeder sizing and know how to compensate for it when sizing conductors.
4. Understanding of voltage drop: Voltage drop is the decrease in voltage along the length of the feeder conductor. You need to understand the importance of minimizing voltage drop in feeder circuits and know how to calculate it for a given feeder circuit.

5. Familiarity with overcurrent protective devices: Overcurrent protective devices (OCPDs), such as circuit breakers and fuses, are used to protect the feeder circuit from overloading and potential damage. You need to be familiar with the different types of OCPDs and their ratings, as well as their role in feeder protection.

Here's an example of how these concepts can be applied in real-world situations: Let's say you're working on a commercial building with a total electrical load of 100 amps. Based on the NEC load calculation requirements, you would determine that a #2 aluminum conductor with a 75-amp ampacity would be appropriate for the feeder circuit. To account for ambient temperature, you would derate the conductor to a 60-amp ampacity, which would then be protected by a 60-amp circuit breaker.

In conclusion, feeder sizing is a critical aspect of electrical design and installation and it is important for you to have a good understanding of the NEC load calculation requirements, conductor ampacity and sizing, the effects of ambient temperature, voltage drop, and overcurrent protective devices for your journeyman electrician exam.

Understanding **voltage drop** is crucial for a journeyman electrician, as it impacts the performance of electrical systems.

Voltage drop is defined as the reduction in voltage in a circuit as a result of electrical resistance and impedance. In other words, as electrical current flows through a conductor, some of the energy is lost as heat due to resistance in the conductor. This results in a reduction in the voltage at the end of the circuit compared to the voltage at the beginning of the circuit.

The National Electrical Code (NEC) provides guidelines for acceptable voltage drop in a circuit. For example, the NEC requires that the voltage drop for feeders and branch circuits does not exceed 3% for feeders and 5% for branch circuits.

To calculate voltage drop, you will need to know the length of the conductor, the wire size, and the current flowing through the conductor. The formula for voltage drop is:
Voltage drop = 2 * Length * Current * Resistance / 1000

Where:
Length = Length of conductor (in feet)
Current = Current flowing through conductor (in amperes)
Resistance = Resistance of conductor (in ohms per 1000 feet)

Here's an example of how to use this formula:
Let's say you have a 200-foot-long conductor with a current flow of 30 amps and a resistance of 0.01 ohms per 1000 feet.
Voltage drop = 2 * 200 * 30 * 0.01 / 1000 = 12 volts

This means that if you have a voltage source of 120 volts, the voltage at the end of the 200-foot-long conductor will be 120 - 12 = 108 volts.

It's important to keep in mind that voltage drop is not the same as power loss, as power loss is a result of both resistance and voltage drop. Nevertheless, voltage drop should still be kept within the limits set by the NEC to ensure that the electrical system operates within safe and acceptable levels.

Protective devices and overcurrent protection are important topics to understand for your journeyman electrician exam, as they play a crucial role in ensuring the safety of electrical systems and the people who use them.

Protective devices, such as circuit breakers and fuses, are designed to interrupt the flow of electrical current in the event of an overload or short circuit. These devices are installed in electrical systems to prevent damage to the system and to protect people from electric shock or electrocution.

Overcurrent protection refers to the protective device and its related circuitry, used to limit the flow of current to a safe level in the event of an overcurrent condition. This is accomplished by monitoring the flow of current and breaking the circuit if the current exceeds a predetermined level.

The National Electrical Code (NEC) provides specific requirements for overcurrent protection, including the size and type of protective device to be used based on the electrical system being protected, as well as the maximum current that can flow through the device. It's important to be familiar with these requirements and to understand how to select the appropriate protective device based on the system requirements.

For example, in a residential electrical system, a 15-amp circuit breaker might be used to protect a general lighting circuit. In this case, the NEC requires that the circuit breaker have a rating of not more than 15 amps, and be capable of interrupting the current in the event of an overcurrent condition.

In real-world examples, consider a scenario where a short circuit occurs in a building. If the overcurrent protection is properly sized and functioning, the circuit breaker will quickly detect the overcurrent condition and interrupt the flow of current, preventing damage to the electrical system and potentially saving lives.

It's also important to note that protective devices and overcurrent protection are not foolproof and regular maintenance is required to ensure they are functioning properly. This includes testing the operation of the protective device and replacing it if it's no longer functioning as intended.

Grounding is an incredibly important topic in the electrical field and is something that every journeyman electrician should be thoroughly familiar with. In this section, we'll go over the basics of grounding, what you need to know for your exam, and how it applies to real-world electrical systems.

Grounding is the process of intentionally connecting a conductor to the earth to establish an electrical connection with the ground. Grounding serves several important purposes in electrical systems, including reducing the risk of electrical shock, minimizing damage to electrical equipment from electrical transients, and providing a path for fault current to flow in the event of a fault.

One of the key things you'll need to know for your journeyman electrician exam is the NEC (National Electrical Code) requirements for grounding. In the NEC, grounding is regulated by a number of different codes, including Articles 250, 300, and 310. Some of the key NEC requirements for grounding include:

All electrical equipment must be grounded.
The grounding electrode system must be capable of conducting fault current.
Grounding conductors must be sized appropriately based on the fault current they are expected to carry.

Grounding conductors must be protected from physical damage.
It's also important to understand the different types of grounding systems that exist, including equipment grounding, system grounding, and grounding electrode systems. Equipment grounding involves connecting electrical equipment to the ground to ensure that any electrical fault will result in a low-impedance path to ground, reducing the risk of electrical shock. System

grounding involves connecting the electrical system to the ground to ensure that electrical potentials are at the same level as the ground potential. Grounding electrode systems involve a number of different components, including grounding electrodes, grounding electrode conductors, and bonding jumpers, that are used to connect the electrical system to the ground.

In terms of real-world examples, consider the grounding system in a commercial building. The electrical system in this building will be connected to the ground through a grounding electrode system, which might consist of a driven rod, a ground ring, or a combination of different types of grounding electrodes. The electrical equipment in the building, such as lighting fixtures, HVAC equipment, and electrical panels, will be connected to the grounding system through equipment grounding conductors.

In terms of sizing, it's important to ensure that the grounding conductors are sized appropriately to carry the maximum fault current that could occur in the electrical system. The NEC provides specific requirements for grounding conductor sizing, based on the size of the electrical service and the type of electrical system being used. For example, the NEC might require a grounding conductor that is #6 copper for a 200A electrical service, or #4 copper for a 400A electrical service.

In conclusion, grounding is a critical component of electrical systems and is essential for ensuring safety, reliability, and performance. Understanding the NEC requirements for grounding, the different types of grounding systems, and how to properly size grounding conductors will be important for your journeyman electrician exam and your career as an electrician.

CHAPTER 3: Branch Circuit Calculations and Conductors.

Branch Circuit Calculations and Conductors is an essential part of electrical theory and principles that is relevant to preparing for your journeyman electrician exam test.

A branch circuit is a specific part of an electrical system that starts at the service panel or distribution board and ends at the electrical device it serves. The electrical current flows through conductors, which are materials that allow electricity to flow through them with minimal resistance.

When calculating branch circuits, it's important to consider the load (the electrical devices that are connected to the circuit) and the conductors (the wire used to carry the electrical current). The conductor size and ampacity (the amount of current a conductor can safely carry) must be sufficient to handle the electrical load. This means that you have to choose the right size and type of conductors to safely carry the electrical current to the devices.

For example, imagine a garden hose as a conductor. Just as a garden hose must be of a certain size to handle a certain amount of water flow, the conductor must be of a certain size to handle a certain amount of electrical current. If the hose is too small, the water flow will be restricted and the hose might burst, just like if the conductor is too small, the electrical current will be restricted and the conductor might overheat and cause a fire.

Conductors are materials that allow electricity to flow through them with ease. This is due to their structure and composition, which allows for the free flow of electrons. Some common examples of conductors include copper, aluminum, and silver. These materials are often used to make electrical wires because they provide a low resistance path for electrical current to flow. Next, let's talk about circuit calculations. These calculations help electricians determine the safe and efficient flow of electrical current in a circuit. There are a number of important calculations that are frequently used in the field, including:

1. Ohm's Law: This law states that the current in a circuit is directly proportional to the voltage across the circuit, and inversely proportional to the resistance of the circuit. The formula for Ohm's Law is $I = V / R$, where I is the current in amps, V is the voltage in volts, and R is the resistance in ohms.
2. Kirchhoff's Laws: There are two Kirchhoff's laws that are important for electricians. The first law, known as Kirchhoff's Voltage Law (KVL), states that the sum of the voltages in a

circuit must equal zero. The second law, known as Kirchhoff's Current Law (KCL), states that the sum of the currents entering and exiting a node in a circuit must be equal.

3. Power: Power is a measure of the rate at which energy is used or generated in a circuit. The formula for power is P = IV, where P is the power in watts, I is the current in amps, and V is the voltage in volts.

4. Resistance: Resistance is a measure of how much a material opposes the flow of electrical current. The formula for resistance is R = V / I, where R is the resistance in ohms, V is the voltage in volts, and I is the current in amps.

These are just a few of the important electrical calculations that you'll need to know for your journeyman electrician exam. It's important to have a solid understanding of these formulas and their applications, so that you can safely and efficiently design and install electrical systems.

Electrical equipment and devices are a vital part of electrical systems and are used to control, distribute, and use electrical power. These include switches, fuses, circuit breakers, transformers, motors, generators, and many more. As a journeyman electrician, you will need to be familiar with these devices and how they work in order to safely and effectively install, maintain, and repair electrical systems.

Let's take a closer look at some of the most common electrical equipment and devices:

1. Switches: A switch is used to turn the flow of electricity on or off. There are several types of switches, including toggle switches, rocker switches, and dimmer switches.

2. Fuses: A fuse is a safety device that protects an electrical circuit from overloading. It contains a metal wire or filament that melts when too much current flows through it, breaking the circuit and preventing damage to the equipment connected to it.

3. Circuit breakers: A circuit breaker is similar to a fuse, but it can be reset after it trips. This makes it a more convenient option for protecting electrical circuits, as you don't have to replace the breaker each time it trips.

4. Transformers: Transformers are used to step up or step down the voltage of an electrical circuit. For example, a transformer may be used to convert high-voltage electricity from a power plant into low-voltage electricity that is safe for use in homes and businesses.

5. Motors: Motors are used to convert electrical energy into mechanical energy. They are found in many appliances, such as fans, pumps, and power tools.

6. Generators: Generators are used to produce electrical energy, typically by converting mechanical energy from a gasoline or diesel engine into electrical energy.

When it comes to preparing for your journeyman electrician exam, it's important to understand the different types of electrical equipment and devices and how they work. You should also be familiar with electrical codes and standards, as well as safety procedures for working with electrical equipment. Additionally, you will need to be able to perform calculations and understand electrical circuits and schematics. With this knowledge and experience, you will be well-prepared for your exam and for your career as a journeyman electrician.

A switch is a device that makes or breaks an electrical circuit, meaning it either completes or interrupts the flow of electrical current in a circuit. It is an important component in electrical systems and plays a crucial role in controlling the flow of electricity.

In your Journeyman electrician exam, you might be asked about different types of switches, such as single-pole switches, double-pole switches, three-way switches, four-way switches, etc.

Single-pole switches are used to control a light or other device from a single location, while double-pole switches are used to control devices that require two separate circuits, such as 240-volt appliances. Three-way switches are used to control a light or other device from two different locations, while four-way switches are used to control the same device from three or more locations.

It is also important to understand the concepts of switch rating, amperage, and voltage. The switch rating is the maximum amount of current the switch can safely handle, which is usually indicated on the switch itself. The amperage is the amount of electrical current flowing through the circuit, and the voltage is the force that drives the current through the circuit.

Another important aspect of switches is switch wiring. You need to understand how to wire different types of switches and how to troubleshoot switch wiring problems.

In terms of real-world examples, switches can be found in various electrical systems, including residential, commercial, and industrial settings. For example, in your home, you might use a single-pole switch to control a light in a room, while in a commercial setting, you might use a double-pole switch to control an HVAC system.

So, these are some of the key concepts related to switches that you should know for your Journeyman electrician exam. It's important to thoroughly understand these concepts and be able to apply them in different scenarios to be successful on your exam.

there are several National Electric Code (NEC) codes related to switches that you should be familiar with when preparing for your journeyman electrician exam. Some of the most important codes include:

1. NEC Article 404.2 - This code covers the installation and use of switches, including the proper placement of switches and their associated wiring.
2. NEC Article 404.3 - This code covers the minimum wire size and ampacity required for switches, as well as the necessary protection against overloading and short circuits.
3. NEC Article 404.4 - This code covers the marking and identification of switches, including the use of permanent and legible labels that clearly indicate the purpose of each switch.
4. NEC Article 404.5 - This code covers the use of switches in wet or damp locations, including the use of switches that are specifically rated for use in such conditions.
5. NEC Article 404.8 - This code covers the use of switches in hazardous (classified) locations, including the use of explosion-proof switches in areas where flammable gases or vapors are present.

Let's talk about **fuses** and their role in electrical systems.

A fuse is a safety device that is designed to protect electrical circuits from damage due to overloading or short-circuits. It works by breaking the circuit if the current flowing through it exceeds a predetermined value. This value is known as the fuse's current rating and is specified in amps.

When a fuse "blows", it's because the heat generated by the high current flowing through the fuse has caused the metal inside the fuse to melt and separate, breaking the circuit. This prevents the excess current from causing damage to the wiring, appliances, or other electrical equipment connected to the circuit.

When it comes to the journeyman electrician exam, it's important to understand the different types of fuses and how they are used in electrical systems. Some common types of fuses include cartridge fuses, plug fuses, and low voltage fuses. Each type of fuse is designed to be used in specific applications and it's important to select the right type of fuse for each circuit to ensure the right level of protection.

Additionally, it's important to be familiar with the National Electric Code (NEC) guidelines for fuse selection and installation. For example, the NEC requires that fuses be installed in an accessible location and that their current ratings be clearly marked.

Real-world examples of the use of fuses include homes, offices, and factories. In homes, fuses are often found in the electrical service panel and protect individual circuits, such as the lighting circuit, the outlets circuit, and the appliances circuit. In factories, fuses are used to protect large electrical systems, such as the motors that drive production equipment.

In conclusion, understanding fuses and their role in electrical systems is an important aspect of preparing for the journeyman electrician exam. Knowing the different types of fuses, how they work, and how to select and install them according to the NEC will help you be successful on the exam.

A **circuit breaker** is a device designed to automatically interrupt electrical flow in a circuit in case of an overload or short circuit. It works by detecting an electrical problem and then quickly turning off the electrical power to prevent damage to the wiring or electrical devices connected to the circuit.

One of the key things you'll need to understand about circuit breakers for your journeyman electrician exam is the National Electric Code (NEC) regulations related to them. The NEC sets standards for the installation and use of circuit breakers to ensure the safety of electrical systems and equipment.

For example, the NEC requires that each circuit breaker have a specific ampacity, which is the amount of electrical current it can handle before tripping. The NEC also sets requirements for the type of circuit breaker that can be used in different situations, such as the use of ground-fault circuit interrupters (GFCIs) in wet or damp locations to prevent electrical shock.

In terms of real-world examples, circuit breakers are used in homes, businesses, and industrial facilities to protect electrical systems and devices from damage. For example, a circuit breaker in your home might trip if you plug in too many appliances at once, or if there's a short circuit in one of the electrical devices. By tripping, the circuit breaker protects your electrical system from damage and also helps prevent electrical fires.

In addition to the NEC regulations, you may also need to understand the different types of circuit breakers, such as standard circuit breakers, ground-fault circuit interrupters (GFCIs), and arc-fault circuit interrupters (AFCIs).

It's also important to understand how to test and reset circuit breakers, and how to safely replace a circuit breaker if necessary. This is a key part of electrical maintenance and troubleshooting, and is something you'll need to be able to do as a journeyman electrician.

Transformers are electrical devices that are used to transfer electrical energy from one circuit to another through electromagnetic induction. This is a critical component in electrical systems and is often tested in electrician certification exams like the Journeyman electrician test.

A transformer works by using two coils of wire, known as the primary and secondary coils. The primary coil is connected to an AC voltage source, and an alternating magnetic field is created in the coil. This magnetic field induces a voltage in the secondary coil, which can then be used to power another electrical circuit.

The basic function of a transformer can be compared to filling a bucket with water from a hose. The water flowing through the hose is like the electrical energy flowing through the primary coil, and the bucket is like the secondary coil. As the water is transferred from the hose to the bucket, it is transformed from one form of energy to another.

When preparing for a Journeyman electrician exam, it's important to understand the specific numbers and calculations associated with transformers. For example, you should be familiar with the formula for calculating the number of turns in a coil, which is given by: $N = (V\_p / V\_s) * (A\_s / A\_p)$, where N is the number of turns, V_p and V_s are the primary and secondary voltages, and A_p and A_s are the cross-sectional areas of the primary and secondary coils.

You should also be familiar with the concept of impedance, as transformers can be used to match the impedance of a load to the impedance of a power source. Additionally, it's important to understand the basic types of transformers, including step-up transformers, step-down transformers, and isolation transformers, and the different applications for each type.

Overall, a solid understanding of transformers and their related calculations and concepts is essential for success on the Journeyman electrician exam and in your future career as an electrician.

Motors are the workhorses of the electrical world and are used in just about every aspect of our lives, from operating fans and pumps to driving conveyor belts and assembly lines. As a Journeyman Electrician, it is important to have a good understanding of how motors work and how to install and maintain them.

So, let's dive in!

A motor is essentially a device that converts electrical energy into mechanical energy. It does this by using an electromagnet to create a rotating magnetic field. This rotating magnetic field then interacts with the permanent magnets in the motor to produce rotational force, or torque. The direction of this torque depends on the direction of the current flowing through the electromagnet.

One of the key things to understand about motors is their efficiency. The efficiency of a motor is a measure of how much mechanical energy it produces relative to the amount of electrical energy it consumes. This is an important factor to consider when selecting a motor for a specific application, as it can greatly impact the operating costs over the life of the motor.

Another important aspect of motors is their starting and running current. The starting current is the amount of current required to get the motor up to speed, while the running current is the amount of current required to keep the motor running at a constant speed. These two values can be very different, with the starting current often being much higher than the running current. This is why many motors are designed with a starting mechanism, such as a capacitor, that helps to reduce the starting current.

When it comes to the Journeyman Electrician exam, you can expect to be tested on your knowledge of the different types of motors, how they work, and how to install and maintain them. You may also be tested on your ability to calculate the size and type of conductor required for a specific motor installation based on the motor's full-load current and other electrical characteristics.

So, to summarize, as a Journeyman Electrician, you should have a good understanding of the following when it comes to motors:

- The basic principles of how motors work
- The different types of motors and their characteristics
- The efficiency of motors
- The starting and running current of motors
- The NEC codes and standards related to motor installations
- How to select and install motors for specific applications
- How to maintain and troubleshoot motors

And that's a broad overview of motors and their importance for the Journeyman Electrician exam.

Generators are an important component in electrical systems. A generator is a device that converts mechanical energy into electrical energy. In other words, it takes something like the spinning of a turbine from a steam engine or a windmill and turns it into electricity that can be used to power our homes and businesses.

As a Journeyman Electrician, it's important to understand the basic principles of how generators work, as well as their applications in different scenarios.

First, let's talk about AC (Alternating Current) and DC (Direct Current) generators. AC generators produce AC power that alternates in direction at a certain frequency (usually 60 Hz in the US), whereas DC generators produce direct current that flows in a single direction.

AC generators are more commonly used in power plants because AC power can be easily transformed to different voltages, making it easier to transport over long distances. However, for most of our day-to-day use, we need DC power, so the AC power produced by the generator must be converted to DC using rectifiers.

Now, let's talk about the main components of a generator. A generator typically consists of a rotor (the rotating part of the generator), a stator (the stationary part), and a set of coils. The rotor is typically attached to a prime mover, like a steam turbine or a gasoline engine, which provides the mechanical energy to turn the rotor. As the rotor turns, it generates a magnetic field, which in turn generates electrical energy in the stator.

There are a number of different types of generators, including synchronous generators and asynchronous generators, but they all operate on the same basic principles.

When it comes to sizing a generator, there are a number of factors to consider, including the power demand of the equipment being powered, the voltage and frequency required, and the type of prime mover being used. It's important to size a generator correctly so that it can handle the load it's designed to support.

In addition to these basic principles, there are a number of safety concerns to keep in mind when working with generators. For example, it's important to ensure that the generator is properly grounded, to reduce the risk of electrical shock, and that it is protected by a circuit breaker or fuse, to reduce the risk of fire in the event of a short circuit.

Understanding the different types of **electric wiring methods and material**s is crucial for any electrician, especially for those preparing for the journeyman electrician exam. Let me explain this in greater detail:

1. Wiring Methods: There are different wiring methods that are allowed by the National Electrical Code (NEC), some of the common ones are:
    *   conduit: metal or plastic tubing used to protect electrical cables.
    *   non-metallic conduit (NM): similar to conduit, but made of plastic material.
    *   armored cable (AC): metal cable covered with metal armor to protect the wires.
    *   metal-clad cable (MC): metal cable covered with a metal sheath.
    *   metal conduit systems: systems that consist of metal conduit and fittings.
2. Wiring Materials: Different types of electrical wiring materials are also regulated by the NEC, including:
    *   Copper: Most commonly used for electrical wiring due to its high conductivity and low resistance.
    *   Aluminum: A less expensive alternative to copper but it has a higher resistance and it is more prone to corrosion.
    *   Conductors: Wires used for carrying electrical current. The most common sizes for residential use are #14, #12, #10 and #8.
    *   Insulation: The material surrounding the conductor that provides electrical insulation and protects from electrical shock. Common types of insulation include PVC, rubber and Teflon.

It's important to know these wiring methods and materials in order to ensure safe and efficient installation of electrical systems.

For your journeyman electrician exam, you should be familiar with the NEC codes and guidelines related to these wiring methods and materials. You may also want to know the required wire size for specific electrical loads, such as motors or appliances.

Electrical equipment and devices refer to the various components and appliances used in electrical systems, such as switches, outlets, lighting fixtures, transformers, motors, and more. As an electrician, it's important to understand the different types of electrical equipment and devices, as well as how to install and maintain them safely and effectively.

Here are some key concepts and codes to keep in mind when it comes to Electrical Equipment and Devices:

1. Equipment Ratings: It's important to be familiar with the ratings of electrical equipment and devices, such as voltage, current, and wattage. These ratings can be found on the equipment itself and should be adhered to when installing and using the equipment.
2. Overcurrent Protection: Electrical equipment and devices must be protected from overcurrents, which can damage the equipment or cause a fire. This can be achieved through the use of fuses or circuit breakers, which are designed to break the circuit in the event of an overcurrent.
3. Grounding: Proper grounding is essential to ensuring the safety of electrical equipment and devices. NEC Article 250 outlines the requirements for grounding, including grounding electrodes, grounding conductors, and grounding connections.
4. GFCI Protection: Ground Fault Circuit Interrupters (GFCIs) are devices designed to protect against electrical shock. They should be installed in areas where electrical devices may come into contact with water, such as kitchens, bathrooms, and outdoor areas.
5. Wiring Methods: Different types of electrical equipment and devices require different wiring methods, such as conduit, cable, or busways. It's important to be familiar with the wiring requirements for each type of equipment and device.
6. Maintenance and Inspection: Electrical equipment and devices should be inspected and maintained regularly to ensure they are functioning properly and safely. This includes cleaning, tightening connections, and replacing damaged equipment.
7. NEC Codes: The National Electrical Code (NEC) provides specific codes and requirements for the installation and use of electrical equipment and devices. It's important to be familiar with the relevant NEC codes for the equipment and devices you are working with.

Real-world examples of electrical equipment and devices include lighting fixtures in homes and commercial buildings, motors and transformers in industrial settings, and switches and outlets in every type of electrical system. It's important to understand the specific requirements and codes for each type of equipment and device you may encounter in your work as an electrician.

In summary, to be prepared for your journeyman electrician exam, it's important to have a strong understanding of the different types of electrical equipment and devices, their ratings, overcurrent protection, grounding requirements, GFCI protection, wiring methods, maintenance and inspection, and relevant NEC codes. Remember to always prioritize safety and follow all codes and requirements when working with electrical equipment and devices.

In the electrical industry, **equipment ratings** refer to the maximum voltage, current, or power that a piece of equipment can safely handle. These ratings are critical to ensuring safe and reliable operation of electrical systems.

For example, a motor or transformer will have a voltage rating, which specifies the maximum voltage it can be safely operated at. If a motor or transformer is operated at a voltage higher than its rated voltage, it may overheat or fail, potentially causing damage or even a safety hazard.

Similarly, circuit breakers, fuses, and other protective devices have current ratings, which specify the maximum current that they can safely interrupt. If a circuit breaker is used on a circuit with a higher current than its rated current, it may not be able to properly interrupt the circuit in the event of a fault, potentially leading to a dangerous electrical situation.

Power ratings are also important, particularly for devices that consume a lot of power, like heaters and air conditioners. These devices have a power rating, which specifies the maximum power they can safely handle. If a device is operated at a power level higher than its rated power, it may overheat or fail, potentially leading to a safety hazard.

NEC provides guidelines for equipment ratings in Article 110. For example, NEC 110.3(A)(1) requires that electrical equipment be installed and used in accordance with its listing and labeling instructions. These instructions will typically include the equipment ratings.

Real-world examples of equipment ratings can be found in almost any electrical system. For example, a 480V transformer may have a voltage rating of 480V, a current rating of 30A, and a power rating of 15kVA. It is important for electricians to understand these ratings to ensure safe and reliable operation of electrical systems.

Grounding is a crucial part of electrical safety, and it is essential to understand the principles and requirements related to grounding in preparation for the journeyman electrician exam.

Grounding is the process of connecting electrical equipment and systems to the earth in order to reduce the risk of electrical shock and fires caused by electrical faults. The National Electrical Code (NEC) provides specific requirements for grounding that must be followed to ensure safe electrical installations.

Some of the key principles to understand when it comes to grounding include the importance of a low-impedance path to the ground, the requirement for all non-current carrying metal parts to be grounded, and the importance of grounding for electrical fault protection.

Real world examples of grounding include installing a grounding rod in the earth and connecting it to the electrical system, using grounding straps to bond equipment to the grounding system, and ensuring that electrical systems are properly grounded to protect against electrical shock and fires.

Specific NEC codes related to grounding include Article 250, which covers grounding and bonding requirements for electrical installations. It is important to study this article in depth to understand the specific requirements related to grounding, including conductor sizing and installation, equipment grounding, and grounding electrode systems.

Overall, it is essential to have a solid understanding of the principles and requirements related to grounding in preparation for the journeyman electrician exam. By studying the NEC and learning from real-world examples, you can gain the knowledge and skills needed to ensure safe and effective electrical installations.

GFCI stands for Ground Fault Circuit Interrupter. It is a device that is designed to protect people from electric shocks caused by ground faults. A ground fault is an electrical fault that occurs when the electrical current deviates from its intended path and flows through an unintended path, such as a person's body. GFCIs work by constantly monitoring the current flow in the circuit and shutting off the power when it detects a ground fault.

GFCIs are required by the NEC in certain locations, such as bathrooms, kitchens, outdoor areas, and unfinished basements. They are also required for receptacles installed within six feet of a sink in laundry and utility rooms.

There are two types of GFCIs: the GFCI receptacle and the GFCI circuit breaker. The GFCI receptacle is installed in the wall, and all electrical devices plugged into it are protected. The GFCI circuit breaker, on the other hand, is installed in the electrical panel and provides GFCI protection to an entire circuit.

It's important to test GFCIs regularly to make sure they are functioning properly. The NEC requires that GFCIs be tested monthly. Testing involves pressing the "test" button on the GFCI, which should cause the power to turn off. If the power does not turn off, the GFCI may be faulty and should be replaced.

In summary, GFCI protection is an essential safety measure in any electrical system, and it is important to be familiar with its installation, operation, and maintenance for the journeyman electrician exam.

Wiring methods are a crucial aspect of electrical installations, as they determine how conductors are routed and protected. There are several wiring methods that can be used, each with its own set of requirements and restrictions. It is essential to follow the National Electric Code (NEC) when installing wiring systems.

One of the most common wiring methods is the conduit system, which involves running wires through metal or plastic conduit that is then installed in walls or ceilings. This method provides excellent protection for the wires and can be used in a wide range of applications. However, it can be difficult to install and may require specialized tools and equipment.

Another wiring method is cable trays, which involve running wires through metal trays that are mounted on walls or ceilings. This method is often used in industrial settings, as it can handle large amounts of wiring and can be easily modified or expanded. However, it may not provide as much protection for the wires as the conduit system.

Raceways are also a popular wiring method, which involve running wires through a plastic or metal channel that is installed on walls or ceilings. This method is often used in commercial buildings, as it provides a clean and professional look. However, it may not provide as much protection for the wires as the conduit system.

In general, it is essential to use the appropriate wiring method for the application and to follow all NEC requirements for the installation. This may involve selecting the appropriate type of wire, properly grounding the system, and following proper installation and termination procedures.

To give a specific example, let's say you are wiring a new residential building. The NEC requires that all wiring be installed in accordance with the NEC Article 300, which outlines the requirements for wiring methods. For example, all wires must be protected from damage and must be supported by approved methods, such as conduit or cable trays.

Another important consideration when wiring a building is the use of ground fault circuit interrupters (GFCIs). These devices are designed to protect against electrical shocks and are required in many applications. For example, the NEC requires GFCIs to be installed in all outdoor outlets, as well as in areas where water may be present, such as bathrooms and kitchens.

In summary, understanding wiring methods and the NEC requirements is essential for any journeyman electrician. It is important to select the appropriate wiring method for the application and to follow all NEC requirements for the installation. Additionally, proper grounding and the use of GFCIs can help ensure the safety of the electrical system.

Chapter 4: Motors and Generators:

Motors and generators are essential components in various electrical systems. A motor is an electric device that converts electrical energy into mechanical energy, while a generator does the opposite, converting mechanical energy into electrical energy. Here are some key points to know about motors and generators:

1. Types of motors: There are several types of motors, including AC motors, DC motors, synchronous motors, and induction motors. AC motors are widely used in industrial and commercial settings, while DC motors are often used in smaller applications like power tools. Synchronous motors are used in applications that require precise control of speed, such as in clock mechanisms. Induction motors are the most commonly used type of motor, and they work by using a rotating magnetic field to turn the rotor.

2. Motor efficiency: Motor efficiency is an important factor to consider when selecting a motor for a specific application. The efficiency of a motor is determined by the ratio of the output power to the input power. The National Electrical Manufacturers Association (NEMA) has set standards for motor efficiency, and the higher the efficiency rating, the more energy-efficient the motor is.

3. Motor protection: Motors must be protected from overloading, overheating, and other hazards that can cause damage or failure. Overload protection devices, such as circuit breakers and fuses, can prevent damage caused by excessive current. Thermal protection devices, such as thermostats and thermal switches, can protect against overheating.

4. Generator types: There are two main types of generators: AC generators and DC generators. AC generators are commonly used in power plants, while DC generators are often used in applications like battery charging.

5. Generator sizing: When selecting a generator, it's important to choose the correct size for the application. The size of a generator is determined by the power output, which is measured in kilowatts (kW) or megawatts (MW). The National Electric Code (NEC) provides guidelines for sizing generators for specific applications.
6. Generator grounding: Generators must be properly grounded to prevent electrical shock hazards. The NEC provides guidelines for grounding and bonding of generators, and failure to comply with these guidelines can result in serious safety hazards.
7. Real-world applications: Motors and generators are used in a wide variety of applications, from powering large industrial machines to providing backup power for homes during a power outage. Understanding the principles of motors and generators is essential for anyone working in the electrical field.

In summary, understanding the types, efficiency, protection, sizing, grounding, and real-world applications of motors and generators is essential for anyone preparing for the journeyman electrician exam.

Motors are devices that convert electrical energy into mechanical energy. They are commonly used in various industrial, commercial, and residential applications. In order to prepare for your journeyman electrician exam, it is important to understand the basic principles of motors, as well as their applications and safety considerations.

There are several types of motors, including AC motors, DC motors, and synchronous motors. AC motors are the most commonly used motors and include induction motors, split-phase motors, and capacitor-start motors. DC motors are used in applications where speed control is important, such as in electric vehicles or robotics. Synchronous motors are used for precision applications, such as in clocks or time-keeping devices.

When selecting a motor, it is important to consider factors such as the motor's voltage and horsepower rating, as well as its starting and running currents. Additionally, it is important to properly install and wire the motor, ensuring that it is connected to the correct voltage and that the wiring is appropriately sized and grounded.

Motor safety is also an important consideration, as motors can be dangerous if not used or maintained properly. Motor safety can be achieved through the use of safety guards, proper grounding and bonding, and adherence to appropriate safety procedures and codes.

As for NEC codes related to motors, there are several sections that address motor installations and safety. For example, Article 430 covers motors, motor circuits, and controllers, while Article 440 covers air conditioning and refrigeration equipment.

Motors are an essential part of electrical systems, and there are several types of motors that you should be familiar with for your journeyman electrician exam. Here are some of the main types of motors and what you need to know about them:

1. AC Motors: AC motors are commonly used in industrial and commercial applications. They use alternating current to power the motor and convert electrical energy into mechanical energy.
2. DC Motors: DC motors are used in a wide range of applications, including electric vehicles and robotics. They use direct current to power the motor and convert electrical energy into mechanical energy.
3. Synchronous Motors: Synchronous motors are used for high-performance applications and are typically used in high-end industrial equipment. They use AC power and have a fixed speed.
4. Induction Motors: Induction motors are commonly used in residential and commercial applications. They use AC power and have a variable speed.
5. Servo Motors: Servo motors are used in robotics, automation, and CNC machines. They are highly precise and can be controlled with great accuracy.

In addition to understanding the different types of motors, it's important to know how to install and maintain them properly. You should be familiar with the National Electric Code (NEC) requirements for motor installation and grounding, as well as the safety precautions that must be taken when working with motors.

For example, NEC Article 430 provides guidelines for motor installation, including the size and type of wire and conduit needed, the type of motor controller required, and the requirements for motor grounding. You should also be familiar with the different types of motor starters and how to select the appropriate one for a given application.

Real-world examples of motors include industrial equipment such as conveyor belts, pumps, and compressors, as well as residential applications such as air conditioning units and refrigerators. By understanding the different types of motors and how they are used in different applications, you will be well-prepared for your journeyman electrician exam.

Motor efficiency is a key consideration in the design and selection of electric motors. It is a measure of the amount of energy that is converted into mechanical power, as compared to the amount of energy that is lost in the form of heat or other losses.

Efficiency is typically expressed as a percentage, and can vary depending on factors such as the motor's design, load, and operating conditions. High-efficiency motors are typically more expensive, but they can provide significant energy savings over their lifetime.

The National Electrical Code (NEC) includes requirements for motor efficiency in certain applications, such as motors used in air-handling systems, pumping systems, and other applications that have a high potential for energy savings.

To maximize motor efficiency, it is important to properly size and select the motor based on the specific application requirements. It is also important to properly maintain and operate the motor, ensuring that it is kept clean, well lubricated, and operating within its designed load range.

In summary, a good understanding of motor efficiency is important for electricians and is a key factor in designing and selecting electric motors for a wide range of applications. Understanding the NEC's requirements for motor efficiency in certain applications is also critical for ensuring compliance with safety and energy efficiency regulations.

Motor protection is an important concept in the electrical industry, especially for electricians working with motors. Motor protection is the process of protecting a motor from damage due to overloading, overheating, short-circuits, and other electrical faults. There are various types of motor protection devices that electricians use, such as overload relays, thermal protectors, fuses, and circuit breakers.

Overload relays are used to protect motors from damage due to overloading, which can cause the motor to overheat and fail. They work by monitoring the motor's current and tripping the motor starter if the current exceeds a certain value for a certain period of time. Thermal protectors are similar to overload relays, but they protect motors from overheating specifically. They contain a thermal switch that opens the circuit if the motor temperature exceeds a certain value.

Fuses and circuit breakers are used to protect motors from short-circuits and other electrical faults. Fuses work by breaking the circuit if the current exceeds a certain value, while circuit breakers trip the circuit when they sense an overcurrent condition.

It is important to select the appropriate motor protection device based on the motor's voltage, horsepower, and other specifications. The National Electrical Code (NEC) provides guidelines for motor protection, including the required size and type of motor protection device.

Real-world examples of motor protection include protecting a conveyor belt motor from overloading, protecting an air conditioning unit motor from overheating, and protecting a water pump motor from short-circuits.

In summary, motor protection is an important concept for electricians working with motors. The various types of motor protection devices, such as overload relays, thermal protectors, fuses, and circuit breakers, are used to protect motors from damage due to overloading, overheating, short-circuits, and other electrical faults. The NEC provides guidelines for motor protection, and it is important to select the appropriate motor protection device based on the motor's specifications.

Generators are electrical devices that convert mechanical energy into electrical energy. They are commonly used in many applications such as backup power for buildings and homes, powering electric vehicles, and as a power source for remote locations where access to the electrical grid is not possible. There are several types of generators, including AC generators, DC generators, synchronous generators, and induction generators.

AC generators, also known as alternators, are commonly used in large-scale power generation applications. They produce electrical energy by rotating a magnetic field around a stationary wire coil. AC generators are the most common type of generator used in power plants and are typically used to produce three-phase power.

DC generators are used to produce direct current (DC) electricity. They work by rotating a magnetic field around a stationary wire coil, but the output of the generator is rectified to produce DC power. DC generators are commonly used in small-scale applications such as electric vehicles and portable power supplies.

Synchronous generators are also used in power plants to produce large amounts of electricity. They are similar to AC generators but are designed to maintain a constant speed and frequency.

Synchronous generators are commonly used in power plants to provide the power needed to run electrical systems, such as lighting and motors.

Induction generators are used to produce electrical energy by using the principle of electromagnetic induction. They are commonly used in small-scale power generation applications such as wind turbines and micro-hydroelectric power plants.

Generator sizing involves selecting a generator that can provide enough power to meet the electrical demands of the intended loads. The size of the generator is typically specified in kilowatts (kW) or megawatts (MW) and is based on the electrical load that it will need to power. There are a few factors to consider when sizing a generator, including the type and number of loads, the starting current of motors and other devices, the voltage, and the duration of the outage.

To determine the appropriate size of a generator, you can perform a load calculation. This involves adding up the power requirements of all the loads that the generator will need to power. You can find the power requirements in the documentation for the equipment or by measuring the current and voltage using a clamp meter and multimeter. Once you know the total power requirement, you can select a generator that is rated for that amount of power, plus some additional capacity to account for starting currents.

The NEC provides guidelines for generator installation, including requirements for grounding, overcurrent protection, and disconnecting means. For example, generators must be grounded and bonded to prevent the buildup of static charges that can create a hazardous condition. The NEC also requires that generators be equipped with overcurrent protection devices to prevent damage to the generator and the equipment it powers.

When it comes to generator grounding, there are a few key things to keep in mind. The first is that generators must be grounded to prevent the build-up of voltage that can occur during operation. This is typically accomplished using a grounding electrode, such as a ground rod or a grounding plate, that is connected to the generator's frame.

It's important to note that generators used as a separately derived system, such as those used to provide emergency power, must have a grounding electrode system that is separate from the building's grounding electrode system. The NEC provides detailed requirements for these grounding systems to ensure that they are properly installed and meet safety standards.

Another important consideration is the use of grounding conductors. These conductors are typically required to be installed in the circuit between the generator and the transfer switch to provide a low-impedance path to ground in the event of a fault. The NEC provides specific requirements for the sizing and installation of these conductors, which vary based on factors such as the size of the generator and the length of the conductors.

It's also worth noting that there are different types of grounding systems that can be used with generators, including solidly grounded, high-resistance grounded, and low-resistance grounded systems. Each of these systems has its own specific requirements and limitations, and it's important to choose the right system for your specific application.

In electrical systems, **control devices and disconnect means** are critical components that are responsible for controlling the flow of electricity, and ensuring safety in the event of an emergency.

Control devices are used to operate and control the function of electrical equipment, while disconnect means are used to safely disconnect electrical equipment from the power supply for maintenance or in case of emergency.

Some examples of control devices include switches, circuit breakers, fuses, and relays. These devices can be used to turn equipment on and off, as well as to protect against overloads or short circuits.

Disconnect means, on the other hand, are typically found in the form of switches or circuit breakers that can be used to disconnect power from specific circuits or equipment. This can be essential in case of an emergency or for routine maintenance.

In order to ensure proper installation and use of these devices, the NEC provides guidelines and standards for their application. For example, section 430.102 of the NEC outlines requirements for motor disconnects, while section 430.109 specifies the use of overload protection devices for motors.

Chapter 5: Special Occupancies, Equipment and Conditions.

Special Occupancies, Equipment and Conditions refer to areas that have unique electrical requirements and may pose additional hazards to people and property. It is important for electricians to be familiar with the different types of special occupancies, equipment, and conditions, as well as the NEC codes that apply to each.

Some examples of special occupancies include hazardous locations like areas where flammable gases or liquids are present, health care facilities, theaters, and places of assembly. Each of these occupancies has specific NEC requirements for the installation and use of electrical equipment.

Electrical equipment can also have unique requirements depending on its intended use. For example, different types of electrical equipment are required for use in wet locations, corrosive environments, or where the temperature is outside of the normal range.

In addition to special occupancies and equipment, there are also special conditions that electricians need to be aware of. These can include emergency systems, standby power systems, and different types of load management systems.

It is important for electricians to be familiar with the NEC codes that apply to these special occupancies, equipment, and conditions. The NEC provides guidance on the design, installation, and use of electrical systems and equipment to ensure safety and compliance with local regulations.

Some specific NEC codes that may apply to special occupancies, equipment, and conditions include Article 500 for Hazardous Locations, Article 517 for Health Care Facilities, and Article 700 for Emergency Systems.

Electrical formula calculations are an essential part of the electrician's job. In Special Occupancies, Equipment and Conditions, you may need to use specific formulas to calculate things such as voltage drop, current, power, and resistance. These formulas can be found in the NEC and are used to ensure that the electrical system is safe and working correctly.

For example, in hazardous locations, you may need to use the formula $I = P/E$ to calculate the current in a circuit. This formula uses the power (P) in watts and the voltage (E) in volts to determine the current (I) in amperes. Similarly, in motor calculations, you may use the formula P

= (I x E x η) / 746 to calculate the power (P) in horsepower, given the current (I) in amperes, voltage (E) in volts, and efficiency (η) of the motor.

It's important to note that in addition to the formulas, you will also need to be familiar with the proper use of measuring devices, such as multimeters and clamp meters, to measure electrical parameters in the field. And always be aware of the NEC codes applicable to the specific Special Occupancies, Equipment and Conditions you are working with.

Understanding electrical formulas and calculations is an essential part of being a journeyman electrician. Some common calculations you should be familiar with include power, voltage, current, resistance, and capacitance calculations.

Power calculations involve using the formula $P = VI$, where P represents power, V represents voltage, and I represents current. For example, if you have a circuit with a voltage of 120 volts and a current of 5 amps, the power would be $P = 120V \times 5A = 600$ watts.

Voltage calculations use the formula $V = IR$, where V represents voltage, I represents current, and R represents resistance. For example, if you have a circuit with a current of 10 amps and a resistance of 5 ohms, the voltage would be $V = 10A \times 5\Omega = 50$ volts.

Current calculations use the formula $I = V/R$, where I represents current, V represents voltage, and R represents resistance. For example, if you have a circuit with a voltage of 120 volts and a resistance of 10 ohms, the current would be $I = 120V/10\Omega = 12$ amps.

Resistance calculations use the formula $R = V/I$, where R represents resistance, V represents voltage, and I represents current. For example, if you have a circuit with a voltage of 240 volts and a current of 20 amps, the resistance would be $R = 240V/20A = 12$ ohms.

Capacitance calculations involve using the formula $C = Q/V$, where C represents capacitance, Q represents charge, and V represents voltage. For example, if you have a circuit with a charge of 10 coulombs and a voltage of 5 volts, the capacitance would be $C = 10C/5V = 2$ farads.

Electrical safety is of utmost importance in any electrical installation, especially when it comes to special occupancies, equipment, and conditions. The National Electrical Code (NEC) sets out specific requirements for electrical safety, and it is important for electricians to understand these requirements to ensure the safety of themselves and others.

One key aspect of electrical safety is grounding. All electrical systems should be grounded to prevent electrical shock and reduce the risk of fire. The NEC requires that all service entrance equipment be grounded and that grounding electrodes be installed to provide a path to ground for fault current.

Another aspect of electrical safety is overcurrent protection. Overcurrent devices, such as fuses and circuit breakers, protect against overloads and short circuits that can lead to fires or damage to equipment. The NEC sets out specific requirements for the sizing and installation of overcurrent protection devices.

In certain special occupancies, such as hazardous locations, additional safety measures must be taken. For example, equipment in hazardous locations must be designed to prevent ignition of flammable materials, and wiring methods must be approved for use in hazardous locations.

It is also important to use personal protective equipment (PPE) when working with electricity. PPE, such as gloves, eye protection, and hearing protection, can prevent injury from electrical shock and reduce exposure to hazardous materials.

In summary, understanding electrical safety requirements is crucial for any electrician, especially when working with special occupancies, equipment, and conditions. The NEC sets out specific requirements for grounding, overcurrent protection, and other safety measures, and it is important for electricians to understand and follow these requirements to ensure a safe electrical installation.

Renewable energy technologies are becoming increasingly popular and important in the world of electricity. As a journeyman electrician, it's important to understand the different types of renewable energy technologies that are commonly used.

One of the most well-known forms of renewable energy is solar power. This involves the use of photovoltaic panels to convert sunlight into electrical energy. The National Electrical Code (NEC) has specific requirements for the installation and wiring of photovoltaic systems, so it's important to be familiar with these codes.

Another popular form of renewable energy is wind power. This involves the use of wind turbines to generate electrical energy. Again, the NEC has specific requirements for the installation and wiring of wind turbine systems, so it's important to be familiar with these codes as well.

Hydroelectric power is another form of renewable energy that involves the use of moving water to generate electrical energy. Geothermal energy, which involves the use of heat from the earth to generate electrical energy, is another form of renewable energy.

It's important to note that while renewable energy technologies offer many benefits, they also have their own unique safety considerations. For example, solar panels can generate high voltages, and wind turbines can have moving parts that pose hazards to workers. As a journeyman electrician, it's important to be familiar with the safety precautions and standards associated with each type of renewable energy technology.

Chapter 6: NEC Codes, Formulas, and Calculations.

As a journeyman electrician, you should have a solid understanding of the National Electrical Code (NEC) and how to use it for calculations and design. Here are a few specific NEC codes and formulas you should be familiar with:

1. Conductor sizing: NEC Article 310 provides guidance on how to size conductors for various applications. The most commonly used formula for calculating conductor size is the ampacity calculation, which takes into account the current carrying capacity of the conductor, the ambient temperature, and the insulation type.

2. Overcurrent protection: NEC Article 240 provides guidelines for selecting the appropriate overcurrent protection for conductors and equipment. The most commonly used formula for sizing overcurrent protection devices (OCPDs) is the ampacity calculation, which involves multiplying the conductor ampacity by a derating factor based on the application and the type of OCPD being used.

3. Voltage drop: NEC Article 210.19 provides guidance on the maximum allowable voltage drop for branch circuits. The most commonly used formula for calculating voltage drop is the voltage drop formula, which involves multiplying the current in the circuit by the resistance of the conductor and the length of the run.

4. Short-circuit current: NEC Article 110.9 provides guidance on the maximum allowable short-circuit current for electrical equipment. The most commonly used formula for calculating short-circuit current is the bolted fault current calculation, which involves determining the available fault current at the service entrance and comparing it to the short-circuit current rating of the equipment being protected.

The National Electrical Code (NEC) is a set of standards and guidelines that govern the installation, use, and maintenance of electrical systems and equipment. It is important for electricians to be familiar with **NEC code requirements** in order to ensure the safety of themselves and others.

Some specific NEC code requirements that are important for electricians to know include:

- Electrical conductor sizing: NEC provides guidelines for determining the appropriate size of electrical conductors based on factors such as the amount of current they will carry, the distance they will travel, and the temperature of the environment in which they will be installed. Electricians should be familiar with the NEC tables and formulas for determining conductor size.

- Grounding and bonding: NEC requires that electrical systems be properly grounded and bonded to prevent the buildup of dangerous electrical charges. Electricians should be familiar with the NEC requirements for grounding and bonding of electrical systems and equipment.

- Wiring methods: NEC provides guidelines for the types of wiring methods that can be used in different environments, such as conduit, cable trays, and raceways. Electricians should be familiar with the NEC requirements for wiring methods.

- Electrical protection: NEC requires that electrical systems be protected from overcurrents and short circuits. Electricians should be familiar with the NEC requirements for electrical protection, including the sizing and selection of overcurrent protective devices.

- Special occupancies and conditions: NEC includes specific requirements for electrical systems in special occupancies and conditions, such as hazardous locations, health care facilities, and swimming pools. Electricians should be familiar with the NEC requirements for these special situations.

It's important for electricians to stay up-to-date with the latest NEC code requirements as they may change over time. Familiarity with NEC code requirements is essential for electricians to ensure the safety of themselves, their colleagues, and the general public.

let's dive into **NEC codes for electrical formulas and calculations**. There are a number of NEC code requirements that are relevant for electrical formulas and calculations.

One important NEC code is Article 220, which covers the calculation of service and feeder loads. This code specifies the methods that should be used for calculating the minimum load requirements for various types of buildings and structures.

Another important NEC code is Article 210, which covers the calculation of branch circuits. This code specifies the maximum ampacity and minimum wire size for branch circuits based on the load and the length of the circuit.

NEC Article 430 covers the calculation of motor loads, which includes requirements for motor starting current, motor horsepower, and overload protection.

Furthermore, NEC Article 250 covers grounding and bonding requirements. This includes sizing the grounding electrode conductor based on the size of the service conductors.

The NEC codes for conductor sizing and protection provide guidelines for selecting the appropriate conductors based on the maximum allowable ampacity and for protecting the conductors from overcurrent.

One of the most important factors to consider when sizing conductors is the maximum allowable ampacity, which is determined by the current-carrying capacity of the conductor and the ambient temperature. NEC Table 310.16 provides ampacity values for conductors based on their size and insulation type, and it also provides correction factors for different ambient temperatures.

To protect conductors from overcurrent, you should use fuses or circuit breakers that are appropriately sized for the conductor. NEC Article 240 provides guidelines for selecting the correct overcurrent protection device based on the conductor size and the application.

It's also important to note that when conductors are run in parallel, the ampacity of each individual conductor is reduced. NEC Section 310.4 provides guidelines for sizing conductors in parallel.

In addition to conductor sizing and protection, the NEC also provides guidelines for grounding and bonding. NEC Article 250 provides information on grounding requirements for electrical systems, while Article 310 provides guidelines for bonding and grounding conductors.

Real world examples of NEC codes for conductor sizing and protection include calculating the minimum wire size and selecting an appropriate overcurrent protection device for a 20-amp lighting circuit in a commercial building.

The NEC has specific codes and requirements that must be followed when installing motors and transformers. One of the main considerations is the sizing of the conductors and overcurrent protection devices. The size of the conductor must be able to handle the load of the motor or transformer, and the overcurrent protection must be sized accordingly.

For example, if you were installing a 10 horsepower motor, you would need to use conductors that are sized for at least 125% of the full load current of the motor. This is outlined in NEC Article 430.22. The overcurrent protection device would also need to be sized to protect the conductors and the motor. The specific code references for overcurrent protection can be found in NEC Article 430.52.

Another important consideration is the installation location of the motor or transformer. Special rules may apply for motors and transformers installed in hazardous locations, such as explosive atmospheres or in areas where flammable liquids or gases may be present. NEC Article 500 covers these requirements.

Overall, it's important to be familiar with NEC Article 430 for motors and NEC Article 450 for transformers, as well as any additional requirements outlined in other applicable articles. It's important to carefully review and follow these codes and requirements to ensure safe and reliable installations.

NEC Codes for Voltage drop: is the decrease in electrical potential that occurs when electricity flows through a wire or cable over a distance. The NEC has specific rules for voltage drop to ensure that electrical equipment and systems work safely and effectively. The code requires that the voltage drop be no more than 5% for branch circuits and 3% for feeders.
To calculate voltage drop, you will need to know the resistance of the wire or cable, the length of the wire or cable, the amperage of the load, and the voltage of the circuit. The formula for voltage drop is:
$Vd = (2 \times L \times R \times I) / 1000$
where:
- Vd is the voltage drop in volts
- L is the length of the wire or cable in feet
- R is the resistance of the wire or cable in ohms per 1000 feet
- I is the amperage of the load in amps

Overcurrent protection is an important aspect of electrical safety. Overcurrents occur when the electrical current exceeds the rating of the conductor or equipment. The NEC has specific rules for overcurrent protection to ensure that electrical systems are safe and reliable.

The code requires overcurrent protection for all conductors and equipment that are not capable of handling the full load current. The code also requires that overcurrent protection be located at the point where the conductor receives its power.

The most common types of overcurrent protection are fuses and circuit breakers. The NEC has specific rules for the sizing of overcurrent protection devices based on the size of the conductor and the type of load. These rules are designed to ensure that the overcurrent protection device will trip before the conductor or equipment is damaged.

let's go over the NEC codes for residential and commercial load calculations. Load calculations are used to determine the amount of power required for a building to operate, and it's important to make sure that the electrical system can handle the load without becoming overloaded. The NEC has specific guidelines for load calculations to ensure safety and efficiency.

For residential load calculations, the NEC provides specific formulas for determining the minimum size of service and feeder conductors, as well as the minimum size of the branch circuit conductors. These formulas take into account factors such as the square footage of the dwelling unit, the number of small appliance branch circuits, and the type of heating and cooling equipment being used. The NEC also provides tables for determining the minimum ampacity of service conductors and the minimum rating of service disconnecting means based on the calculated load.

For commercial load calculations, the NEC requires that the connected loads be determined, including the general lighting load, the general-use receptacle load, and the load for special equipment. These loads are then converted to demand loads using specific NEC formulas, and the demand loads are then used to determine the minimum size of service and feeder conductors. The NEC also has guidelines for load diversity, which takes into account the likelihood that not all loads will be operating at their maximum at the same time.

It's important to note that these load calculations are just a starting point, and the actual load may be different based on specific circumstances. It's always a good idea to consult the NEC and a licensed electrician to ensure that load calculations are done correctly and safely.

To prevent page flipping, we have done a format different from some practice tests. Rather than flipping to an answer key in the back, the answer and explanation will immediately follow the question, again to prevent page flipping. To prevent spoiling the answer, take a spare sheet of paper or something similar to hide the answer. We have found that many students prefer this format instead of flipping through to find the answer key and flipping back and forth.

Please also note that some of the most important topics will be covered several times and some questions will be near repeats of other questions for the ones that are most crucial for you to know for the exam. It may seem repetitive at times, but this is to hammer the most important parts and topics for your exam.

There will be a wide variety of questions, ranging from easy to hard, in no particular order. Good Luck!

What is the electrical term used to describe the resistance to the flow of electrons in a circuit?
a) Voltage
b) Current
c) Capacitance
d) Resistance

Answer: d) Resistance
Explanation: Resistance is the measure of opposition to the flow of electric current. It is measured in ohms (Ω).

What is the electrical term used to describe the rate of flow of electric charge in a circuit?
a) Voltage
b) Current
c) Capacitance
d) Resistance

Answer: b) Current

Explanation: Current is the rate at which electric charge flows in a circuit. It is measured in amperes (A).

What is the electrical term used to describe the amount of electric potential energy per unit charge in a circuit?
a) Voltage
b) Current
c) Capacitance
d) Resistance

Answer: a) Voltage

Explanation: Voltage, also known as electric potential difference, is the amount of electric potential energy per unit charge in a circuit. It is measured in volts (V).

What is the unit of power in the electrical system?
a) Watt
b) Joule
c) Ampere
d) Volt

Answer: a) Watt

Explanation: Power is the rate at which work is done or energy is transferred. In the electrical system, power is measured in watts (W).

What is the electrical term used to describe the ability of a material to store electric charge?
a) Voltage
b) Current
c) Capacitance
d) Resistance

Answer: c) Capacitance

Explanation: Capacitance is the ability of a material to store electric charge. It is measured in farads (F).

Which of the following is true about resistance in a circuit?
a) Resistance decreases with increasing temperature
b) Resistance increases with decreasing temperature
c) Resistance is independent of temperature
d) Resistance can only be measured with an oscilloscope

Answer: b) Resistance increases with decreasing temperature
Explanation: This is because as the temperature decreases, the atoms in the material have less kinetic energy, causing them to vibrate less and obstruct the flow of electrons more, increasing the resistance.

What is the unit of measure for electrical power?
a) Volts
b) Amps
c) Watts
d) Ohms

Answer: c) Watts
Explanation: Watts are the unit of measure for electrical power, defined as the rate at which energy is transferred or used.

What is the purpose of a fuse or circuit breaker in a circuit?
a) To increase the voltage of the circuit
b) To decrease the resistance of the circuit
c) To limit the current in the circuit
d) To add more components to the circuit

Answer: c) To limit the current in the circuit
Explanation: Fuses and circuit breakers are used to protect the circuit and prevent damage to the components by limiting the current in the circuit.

Which of the following is a type of electrical conductor?
a) Glass
b) Rubber
c) Copper
d) PVC

Answer: c) Copper
Explanation: Copper is a common electrical conductor due to its high conductivity and low cost.

What is the direction of current flow in a circuit?
a) From positive to negative
b) From negative to positive
c) In both directions simultaneously
d) It depends on the type of circuit

Answer: b) From negative to positive
Explanation: The conventional direction of current flow is from the negative terminal of the source to the positive terminal. However, in reality, the flow of electrons is actually in the opposite direction, from the negative side of the source to the positive side.

What is voltage?
a) The flow of electric current
b) The resistance to electric current
c) The amount of electric power consumed
d) The potential difference between two points in a circuit

Answer: d
Explanation: Voltage is the potential difference between two points in an electrical circuit. It is measured in volts.

Which of the following is the most common unit of voltage?

a) Watts

b) Ohms

c) Volts

d) Amperes

Answer: c

Explanation: Volts are the most common unit of voltage, and they are used to measure the electrical potential difference between two points in a circuit.

What happens to voltage when resistance is increased in a circuit?

a) Voltage increases

b) Voltage decreases

c) Voltage remains the same

d) Voltage becomes zero

Answer: b

Explanation: When resistance is increased in a circuit, voltage decreases because there is more opposition to the flow of current.

What is the relationship between voltage and current in a circuit?

a) They are directly proportional

b) They are inversely proportional

c) They are unrelated

d) They are equal

Answer: a

Explanation: Voltage and current are directly proportional in a circuit. If voltage increases, then current increases, and if voltage decreases, then current decreases.

What is the voltage of a typical household outlet in the United States?

a) 120 volts
b) 240 volts
c) 480 volts
d) 12 volts

Answer: a

Explanation: A typical household outlet in the United States has a voltage of 120 volts, which is enough to power most household appliances and electronics.

Which of the following is the unit of measurement for electric current?
a) Volt
b) Ampere
c) Ohm
d) Watt

Answer: b) Ampere

Explanation: The unit of measurement for electric current is ampere, which is often shortened to "amp".

What is the relationship between amperes and watts in a circuit?
a) Amperes represent voltage, while watts represent current flow.
b) Amperes represent current flow, while watts represent power.
c) Amperes and watts are unrelated measurements in a circuit.
d) Amperes and watts both measure voltage.

Answer: b) Amperes represent current flow, while watts represent power.

Explanation: Amperes measure the amount of current flowing in a circuit, while watts measure the power being used by the circuit.

What happens when the amperage in a circuit exceeds the capacity of the wires or components?
a) The circuit will become more efficient.
b) The wires or components will heat up and possibly cause a fire.
c) The circuit will shut down automatically to prevent damage.
d) The circuit will produce more voltage.

Answer: b) The wires or components will heat up and possibly cause a fire.

Explanation: When the amperage in a circuit exceeds the capacity of the wires or components, the wires or components will heat up due to the increased resistance, and may eventually overheat and cause a fire.

What is the typical amperage of a standard household outlet in North America?
a) 10 amps
b) 15 amps
c) 20 amps
d) 30 amps

Answer: b) 15 amps

Explanation: A standard household outlet in North America is typically rated for a maximum of 15 amps.

How can the amperage in a circuit be calculated?
a) By dividing voltage by resistance.
b) By multiplying voltage by resistance.
c) By dividing power by voltage.
d) By dividing power by resistance.

Answer: d) By dividing power by resistance.

Explanation: The amperage in a circuit can be calculated using Ohm's law, which states that amperage = power / resistance.

What is the unit of measurement for resistance?
a) Amps
b) Watts
c) Ohms
d) Volts

Answer: c) Ohms
Explanation: The unit of measurement for resistance is Ohms, which is represented by the Greek letter Omega (Ω).

Which of the following factors does NOT affect the resistance of a wire?
a) The length of the wire
b) The diameter of the wire
c) The temperature of the wire
d) The color of the wire

Answer: d) The color of the wire
Explanation: The color of a wire is used to indicate its purpose, but it does not affect its resistance. The length, diameter, and temperature of a wire can all impact its resistance.

What is the formula for calculating resistance in a circuit?
a) R = I x V
b) R = V / I
c) R = P / I^2
d) R = V x I

Answer: b) R = V / I
Explanation: The formula for calculating resistance in a circuit is R = V / I, where R is resistance in Ohms, V is voltage in Volts, and I is current in Amperes.

Which of the following materials has the highest electrical resistance?
a) Copper

b) Silver
c) Aluminum
d) Nichrome

Answer: d) Nichrome
Explanation: Nichrome is a type of alloy that is commonly used in heating elements due to its high electrical resistance. Copper and silver are both good conductors of electricity, while aluminum has a relatively low resistance compared to Nichrome.

If two resistors are connected in series, how does their combined resistance compare to the resistance of each individual resistor?
a) It is equal to the sum of their individual resistances
b) It is equal to the product of their individual resistances
c) It is less than the resistance of each individual resistor
d) It is greater than the resistance of each individual resistor

Answer: a) It is equal to the sum of their individual resistances
Explanation: When resistors are connected in series, their resistances add together to determine the total resistance of the circuit. Therefore, the combined resistance of two resistors connected in series is equal to the sum of their individual resistances.

What is the unit of measurement for electrical resistance?
a) Volts
b) Amperes
c) Ohms
d) Watts

Answer: c) Ohms
Explanation: Electrical resistance is measured in ohms, which is the unit of resistance in the International System of Units (SI).

What is the purpose of a fuse in an electrical circuit?
a) To limit the amount of current in the circuit

b) To increase the voltage in the circuit
c) To measure the resistance of the circuit
d) To switch the circuit on and off

Answer: a) To limit the amount of current in the circuit
Explanation: A fuse is a safety device that is designed to protect an electrical circuit from excessive current by breaking the circuit when the current exceeds a predetermined level.

What is the difference between a series and a parallel circuit?
a) The voltage in a series circuit is constant, while the voltage in a parallel circuit varies.
b) The resistance in a series circuit is constant, while the resistance in a parallel circuit varies.
c) The current in a series circuit is constant, while the current in a parallel circuit varies.
d) The power in a series circuit is constant, while the power in a parallel circuit varies.

Answer: c) The current in a series circuit is constant, while the current in a parallel circuit varies.
Explanation: In a series circuit, the components are connected end-to-end, and the current is the same through each component. In a parallel circuit, the components are connected side-by-side, and the current is divided among the components.

What is the purpose of a circuit breaker in an electrical circuit?
a) To increase the voltage in the circuit
b) To measure the resistance of the circuit
c) To switch the circuit on and off
d) To protect the circuit from excessive current

Answer: d) To protect the circuit from excessive current
Explanation: A circuit breaker is a safety device that is designed to protect an electrical circuit from excessive current by automatically switching off the circuit when the current exceeds a predetermined level

What is the purpose of a fuse in an electrical circuit?
a) To increase voltage
b) To reduce current
c) To protect the circuit from overloading

d) To reduce resistance

Answer: c
Explanation: A fuse is designed to protect an electrical circuit from overloading by opening the circuit and interrupting the flow of current when the current exceeds a safe level.

What is the difference between a series and parallel circuit?
a) A series circuit has only one path for current, while a parallel circuit has multiple paths for current.
b) A series circuit has multiple paths for current, while a parallel circuit has only one path for current.
c) A series circuit has a constant voltage across all components, while a parallel circuit has a constant current across all components.
d) A series circuit has a constant current across all components, while a parallel circuit has a constant voltage across all components.

Answer: a
Explanation: In a series circuit, all components are connected in a single path, so the current flowing through each component is the same. In a parallel circuit, the components are connected in multiple paths, so the current can flow through each component independently.

What is the formula for calculating power in an electrical circuit?
a) Power = Voltage x Current
b) Power = Voltage / Current
c) Power = Current x Resistance
d) Power = Voltage x Resistance

Answer: a
Explanation: The formula for power in an electrical circuit is P = VI, where P is power in watts, V is voltage in volts, and I is current in amperes.

What is the purpose of a ground fault circuit interrupter (GFCI)?
a) To protect the circuit from overloading
b) To reduce voltage
c) To reduce resistance

d) To protect against electric shock

Answer: d
Explanation: A GFCI is designed to protect against electric shock by detecting imbalances in the electrical current and interrupting the circuit.

Which type of circuit breaker is commonly used in residential and commercial electrical systems?
a) Thermal-magnetic circuit breaker
b) Ground fault circuit interrupter (GFCI)
c) Arc fault circuit interrupter (AFCI)
d) Electronic circuit breaker

Answer: a
Explanation: Thermal-magnetic circuit breakers are commonly used in residential and commercial electrical systems to protect against overloading and short circuits. GFCIs and AFCIs are also commonly used in specific applications, such as outdoor outlets and bedrooms, respectively. Electronic circuit breakers are less common and generally used in more specialized applications.

What is capacitance?
a) The property of a circuit to oppose changes in current flow
b) The property of a circuit to store electrical energy
c) The ability of a circuit to conduct electricity
d) The ability of a circuit to resist the flow of electricity

Answer: b) The property of a circuit to store electrical energy

Explanation: Capacitance is the ability of a circuit to store electrical energy in an electric field between two conductive surfaces.

Which of the following factors affects the capacitance of a capacitor?
a) The distance between the plates
b) The size of the plates
c) The material of the plates

d) All of the above

Answer: d) All of the above

Explanation: The capacitance of a capacitor is directly proportional to the size of the plates, the distance between the plates, and the dielectric constant of the material between the plates.

What happens to the capacitance of a capacitor if the distance between the plates is increased?
a) It decreases
b) It increases
c) It stays the same
d) It depends on the size of the plates

Answer: a) It decreases

Explanation: Capacitance is inversely proportional to the distance between the plates. As the distance between the plates increases, the capacitance decreases.

What unit is used to measure capacitance?
a) Ohms
b) Farads
c) Hertz
d) Amperes

Answer: b) Farads

Explanation: Capacitance is measured in farads, named after Michael Faraday, who discovered the principle of capacitance.

What is a capacitor used for in an electrical circuit?
a) To generate a magnetic field
b) To measure the amount of current in a circuit
c) To store electrical energy
d) To regulate voltage

Answer: c) To store electrical energy

Explanation: Capacitors are used in electrical circuits to store electrical energy and to provide a temporary source of power to devices. They are often used in filters, timers, and oscillators.

Which of the following is the formula for calculating power in watts (W)?
a) P = V x I
b) P = I^2 x R
c) P = V^2 / R
d) P = I x R

Answer: c) P = V^2 / R

Explanation: Power (P) is equal to the voltage (V) squared divided by the resistance (R). This formula is known as the power law.

A 120V circuit has a total resistance of 10 ohms. What is the total power (in watts) consumed by the circuit?
a) 12W
b) 144W
c) 1200W
d) 1440W

Answer: b) 144W

Explanation: Using the power law, we can find the power consumed by the circuit by using the formula P = V^2 / R. Substituting the given values, we get P = (120V)^2 / 10 ohms = 1440W. Therefore, the total power consumed by the circuit is 144W.

What is the unit of power?
a) Amps (A)

b) Volts (V)

c) Ohms (Ω)

d) Watts (W)

Answer: d) Watts (W)

Explanation: Power is the rate at which work is done, or energy is transferred per unit of time. It is measured in watts (W).

What is the power factor of a purely resistive circuit?

a) 0.5

b) 1.0

c) 0.0

d) 0.707

Answer: b) 1.0

Explanation: A purely resistive circuit has a power factor of 1.0, which means that the current is in phase with the voltage and there is no reactive power.

What is the difference between real power and reactive power?

a) Real power is the power used by the circuit to perform useful work, while reactive power is the power used by the circuit to generate magnetic fields.

b) Real power is the power consumed by the circuit, while reactive power is the power generated by the circuit.

c) Real power is the power consumed by the circuit to generate magnetic fields, while reactive power is the power used by the circuit to perform useful work.

d) Real power and reactive power are the same thing.

Answer: a) Real power is the power used by the circuit to perform useful work, while reactive power is the power used by the circuit to generate magnetic fields.

Explanation: Real power is the power consumed by the circuit to perform useful work, such as powering a light bulb or a motor. Reactive power, on the other hand, is the power used by the circuit to generate magnetic fields in inductive components such as transformers or motors.

If the resistance of a circuit is 10 ohms and the current flowing through it is 2 amperes, what is the voltage across the circuit?

a) 2 volts

b) 5 volts

c) 10 volts

d) 20 volts

Answer: d) 20 volts

Explanation: Using Ohm's Law, we can calculate the voltage across the circuit as $V = IR = (2 \text{ A}) \times (10 \text{ }\Omega) = 20 \text{ V}$.

A circuit has a resistance of 50 ohms and a voltage of 100 volts. What is the current flowing through the circuit?

a) 0.5 amperes

b) 2 amperes

c) 5 amperes

d) 50 amperes

Answer: a) 0.5 amperes

Explanation: Using Ohm's Law, we can calculate the current flowing through the circuit as $I = V/R = (100 \text{ V}) / (50 \text{ }\Omega) = 2 \text{ A}$.

What happens to the current in a circuit if the voltage is doubled while the resistance remains the same?

a) The current doubles

b) The current is halved

c) The current quadruples

d) The current remains the same

Answer: a) The current doubles

Explanation: Ohm's Law states that the current flowing through a circuit is directly proportional to the voltage, and inversely proportional to the resistance. Therefore, if the voltage is doubled while the resistance remains the same, the current must also double.

What is the resistance of a circuit if a voltage of 12 volts is applied and a current of 3 amperes flows through it?
a) 4 ohms
b) 9 ohms
c) 12 ohms
d) 36 ohms

Answer: a) 4 ohms

Explanation: Using Ohm's Law, we can calculate the resistance of the circuit as $R = V/I = (12 \text{ V}) / (3 \text{ A}) = 4 \text{ }\Omega$.

Which of the following best defines energy?
a) The movement of charged particles
b) The ability to do work
c) The amount of heat generated by an electrical system
d) The resistance of a material to the flow of electric current

Answer: b
Explanation: Energy is the ability to do work, which can include producing light, heat, or mechanical motion.

What unit is used to measure energy?
a) Amperes
b) Volts
c) Watts
d) Ohms

Answer: c

Explanation: Energy is measured in watts (W), which is equal to one joule (J) of energy per second.

Which type of energy is stored in a battery?
a) Mechanical energy
b) Chemical energy
c) Electrical energy
d) Thermal energy

Answer: b

Explanation: A battery stores chemical energy, which can be converted into electrical energy when the battery is connected to a circuit.

What law of thermodynamics states that energy cannot be created or destroyed, only transferred or converted?
a) The first law of thermodynamics
b) The second law of thermodynamics
c) The third law of thermodynamics
d) The zeroth law of thermodynamics

Answer: a

Explanation: The first law of thermodynamics, also known as the law of conservation of energy, states that the total energy in a closed system is constant and cannot be created or destroyed, only transferred or converted from one form to another.

Which of the following is an example of potential energy?
a) A moving car
b) A light bulb turned on
c) A compressed spring
d) A person walking

Answer: c

Explanation: Potential energy is energy that is stored in an object due to its position or configuration, such as a compressed spring or a raised weight.

In a circuit with multiple current paths, what does Kirchhoff's Current Law (KCL) state?
a) The voltage across a series circuit is equal to the sum of the individual voltage drops.
b) The total resistance in a parallel circuit is equal to the sum of the individual resistances.
c) The sum of the currents entering a node equals the sum of the currents leaving the node.
d) The total power output in a circuit is equal to the product of the voltage and the current.

Answer: c) The sum of the currents entering a node equals the sum of the currents leaving the node.

Explanation: KCL states that the total current entering a node or junction in a circuit must equal the total current leaving that node or junction. This is based on the principle of conservation of charge.

In a circuit with multiple voltage sources, what does Kirchhoff's Voltage Law (KVL) state?
a) The total power output in a circuit is equal to the product of the voltage and the current.
b) The voltage across a series circuit is equal to the sum of the individual voltage drops.
c) The voltage across a parallel circuit is equal to the sum of the individual voltages.
d) The sum of the voltage drops around a closed loop in a circuit is equal to the sum of the voltage sources in that loop.

Answer: d) The sum of the voltage drops around a closed loop in a circuit is equal to the sum of the voltage sources in that loop.

Explanation: KVL states that the sum of the voltage drops around any closed loop in a circuit must be equal to the sum of the voltage sources in that loop. This is based on the principle of conservation of energy.

In a circuit with two voltage sources connected in series, what is the voltage drop across each source?
a) The voltage drop is split equally between the two sources.
b) The voltage drop is greater across the source with the larger internal resistance.
c) The voltage drop is greater across the source with the smaller internal resistance.

d) The voltage drop is equal to the voltage of each source.

Answer: d) The voltage drop is equal to the voltage of each source.

Explanation: When voltage sources are connected in series, the total voltage is equal to the sum of the individual voltages. Therefore, each source will have the same voltage drop across it.

In a circuit with two resistors connected in parallel, what is the total resistance of the circuit?
a) The total resistance is equal to the sum of the individual resistances.
b) The total resistance is equal to the product of the individual resistances.
c) The total resistance is less than the smallest individual resistance.
d) The total resistance is greater than the largest individual resistance.

Answer: c) The total resistance is less than the smallest individual resistance.

Explanation: When resistors are connected in parallel, the total resistance is less than the smallest individual resistance. This is because the total current is divided between the resistors, resulting in a lower overall resistance.

In a circuit with a single voltage source and two resistors connected in series, what is the voltage drop across each resistor?
a) The voltage drop is equal across both resistors.
b) The voltage drop is greater across the resistor with the smaller resistance.
c) The voltage drop is greater across the resistor with the larger resistance.
d) The voltage drop depends on the type of resistor used.

Answer: c) The voltage drop is greater across the resistor with the larger resistance.

Explanation: When resistors are connected in series, the voltage drop is divided between them based on their individual resistances.

Which of the following is the correct formula for calculating impedance in an AC circuit?
a) $Z = R + XL$
b) $Z = R + XC$

c) $Z = \sqrt{(R^2 + XL^2)}$
d) $Z = \sqrt{(R^2 + XC^2)}$

Answer: c) $Z = \sqrt{(R^2 + XL^2)}$

Explanation: Impedance (Z) in an AC circuit is the total opposition to current flow, and is calculated using the formula $Z = \sqrt{(R^2 + XL^2)}$, where R is the resistance and XL is the inductive reactance.

If the voltage in a circuit is 120 volts and the impedance is 10 ohms, what is the current in the circuit?
a) 0.12 A
b) 1.2 A
c) 12 A
d) 120 A

Answer: b) 1.2 A

Explanation: Ohm's law states that current (I) = voltage (V) / impedance (Z). Plugging in the values, we get I = 120 V / 10 Ω = 12 A.

Which of the following is a unit of impedance?
a) Volts
b) Amperes
c) Ohms
d) Farads

Answer: c) Ohms

Explanation: Impedance, like resistance, is measured in ohms.

In a series AC circuit, if the resistance is 20 ohms and the inductive reactance is 30 ohms, what is the total impedance of the circuit?

a) 10 ohms

b) 20 ohms

c) 30 ohms

d) 40 ohms

Answer: d) 40 ohms

Explanation: In a series AC circuit, the total impedance (Z) is the sum of the resistance (R) and the inductive reactance (XL). So, $Z = R + XL = 20\,\Omega + 30\,\Omega = 40\,\Omega$.

What happens to the impedance of a circuit as the frequency of the AC voltage increases?

a) Increases

b) Decreases

c) Remains constant

d) Impedance is not affected by frequency

Answer: a) Increases

Explanation: Impedance is a function of frequency in AC circuits, and it generally increases as the frequency increases. This is because higher frequencies result in greater reactance in capacitive and inductive elements, which add to the overall impedance of the circuit.

What is the minimum vertical clearance required over residential service conductors from the roof of a building?

a. 6 ft

b. 8 ft

c. 10 ft

d. 12 ft

Answer: b

Explanation: According to NEC 230.24, the minimum vertical clearance over residential service conductors is 8 ft.

What is the maximum height that a meter socket can be installed on a residential building?
a. 4 ft
b. 6 ft
c. 8 ft
d. 10 ft

Answer: c

Explanation: According to NEC 230.70(A)(1), the bottom of the meter socket must be no higher than 6 ft above the ground or working platform, unless local conditions require a higher installation.

What is the maximum overcurrent protection device (OCPD) size allowed for a 200-amp residential service?
a. 200 A
b. 225 A
c. 250 A
d. 300 A

Answer: c

Explanation: According to NEC 230.90, the maximum OCPD size for a 200-amp residential service is 250 A.

What is the maximum height that a service disconnecting means can be installed on a commercial building?
a. 6 ft

b. 8 ft

c. 10 ft

d. 12 ft

Answer: b

Explanation: According to NEC 230.70(A)(2), the bottom of the service disconnecting means must be no higher than 6 ft above the floor or working platform.

What is the minimum wire size required for a 400-amp commercial service?

a. 2/0 AWG

b. 4/0 AWG

c. 250 kcmil

d. 350 kcmil

Answer: c

Explanation: According to NEC Table 310.16, the minimum wire size required for a 400-amp service is 250 kcmil.

What is the minimum clearance required from the top of the electrical service mast to a balcony or deck?

a) 3 feet

b) 6 feet

c) 9 feet

d) 12 feet

Answer: b

Explanation: According to NEC, the minimum clearance required from the top of the electrical service mast to a balcony or deck is 6 feet.

What is the maximum height for the electrical service meter base above the ground?

a) 5 feet

b) 6 feet

c) 7 feet

d) 8 feet

What is the minimum distance required between the electrical service mast and a window or opening that can be opened?
a) 2 feet
b) 3 feet
c) 4 feet
d) 5 feet

What is the maximum allowable distance between the electrical service entrance and the main service panel?
a) 20 feet
b) 30 feet
c) 40 feet
d) 50 feet

What is the minimum clearance required between the electrical service mast and the edge of a roof that is used for pedestrian traffic?
a) 3 feet
b) 4 feet
c) 5 feet
d) 6 feet

Answer: c

Explanation: According to NEC, the minimum clearance required between the electrical service mast and the edge of a roof that is used for pedestrian traffic is 5 feet.

Which of the following is not a type of electrical distribution system?

a. Overhead

b. Underground

c. Radial

d. Square

Answer: d

Explanation: Radial, overhead, and underground are all common types of electrical distribution systems, but there is no such thing as a "square" distribution system.

What is the purpose of a distribution transformer in an electrical distribution system?

a. To increase voltage

b. To decrease voltage

c. To maintain voltage

d. To produce electricity

Answer: b

Explanation: The distribution transformer is responsible for stepping down the voltage from the transmission level (typically 69 kV or higher) to the distribution level (typically 12 kV or lower) so that it can be used by homes and businesses.

Which type of circuit breaker is typically used in a residential electrical distribution panel?

a. Low-voltage circuit breaker

b. Medium-voltage circuit breaker

c. High-voltage circuit breaker

d. Miniature circuit breaker

Answer: d

Explanation: Miniature circuit breakers (MCBs) are commonly used in residential electrical distribution panels to protect circuits from overloads and short circuits.

In a three-phase electrical distribution system, how many wires are typically used to transmit power?

a. One
b. Two
c. Three
d. Four

Answer: c

Explanation: Three-phase power uses three wires to transmit power, typically with a voltage of 208 V, 240 V, or 480 V.

What is a common method for protecting a distribution system from lightning strikes?

a. Installing surge protectors at each electrical panel
b. Grounding the electrical system
c. Burying the electrical cables
d. Painting the electrical equipment

Answer: b

Explanation: Grounding the electrical system is a common method for protecting against lightning strikes by providing a path for the electrical current to flow safely to the ground.

In commercial buildings, what type of electrical distribution system is most commonly used?

a) Delta system
b) Wye system
c) T system
d) Zigzag system

Answer: b) Wye system

Explanation: The Wye system is the most commonly used electrical distribution system in commercial buildings.

What type of electrical panel is typically used in residential buildings?

a) Circuit breaker panel

b) Fuse panel

c) Switch panel

d) Main breaker panel

Answer: a) Circuit breaker panel

Explanation: Circuit breaker panels are the most common type of electrical panel used in residential buildings, providing protection for individual circuits.

What is the maximum current carrying capacity of a 12 AWG wire in a residential circuit?

a) 15 amps

b) 20 amps

c) 30 amps

d) 40 amps

Answer: b) 20 amps

Explanation: A 12 AWG wire is typically used in residential circuits and has a maximum current carrying capacity of 20 amps.

Which type of electrical conduit is commonly used in commercial buildings?

a) PVC conduit

b) EMT conduit

c) Flexible metal conduit

d) Rigid metal conduit

Answer: d) Rigid metal conduit

Explanation: Rigid metal conduit is commonly used in commercial buildings to protect electrical wiring from damage.

What is the purpose of a GFCI outlet?

a) To protect against overloading circuits

b) To protect against short circuits

c) To protect against electrical shock

d) To protect against power surges

Answer: c) To protect against electrical shock

Explanation: A GFCI (Ground Fault Circuit Interrupter) outlet is designed to detect and interrupt electrical current if it is flowing through an unintended path, such as through a person, in order to protect against electrical shock.

What is a separately derived system?

a) An electrical system that is supplied by a transformer that has one or more ungrounded conductors

b) An electrical system that is supplied by a transformer that has both ungrounded and grounded conductors

c) An electrical system that is supplied by a generator

d) An electrical system that is supplied by a battery

Answer: a) An electrical system that is supplied by a transformer that has one or more ungrounded conductors.

Explanation: A separately derived system is a type of electrical system that is supplied by a transformer that has one or more ungrounded conductors. These conductors are not grounded at the transformer, and the system is grounded at a separate point.

What is the purpose of a separately derived system?

a) To provide a grounding point for electrical equipment and reduce electrical hazards

b) To increase the efficiency of electrical systems

c) To reduce the cost of electrical installations

d) To provide a backup power source in case of power outages

Answer: a) To provide a grounding point for electrical equipment and reduce electrical hazards.

Explanation: The purpose of a separately derived system is to provide a grounding point for electrical equipment and reduce electrical hazards. By grounding the system at a separate point, faults and electrical hazards can be safely isolated and addressed.

Which of the following is an example of a separately derived system?
a) A residential electrical system
b) An industrial electrical system
c) An emergency generator system
d) A battery backup system

Answer: c) An emergency generator system.

Explanation: An emergency generator system is an example of a separately derived system, as it is supplied by a transformer that has one or more ungrounded conductors.

In a separately derived system, what is the purpose of the grounding electrode conductor?
a) To connect the grounding electrode to the transformer
b) To connect the neutral conductor to the transformer
c) To connect the transformer to the load
d) To connect the ungrounded conductor to the load

Answer: a) To connect the grounding electrode to the transformer.

Explanation: The grounding electrode conductor in a separately derived system is used to connect the grounding electrode to the transformer. This provides a separate grounding point for the system, which helps to reduce electrical hazards.

What is the maximum resistance allowed for the grounding electrode in a separately derived system?
a) 5 ohms
b) 10 ohms
c) 25 ohms
d) 100 ohms

Answer: b) 10 ohms.

Explanation: In a separately derived system, the maximum resistance allowed for the grounding electrode is 10 ohms. This ensures that the system is properly grounded and reduces the risk of electrical hazards.

What is the purpose of an electrical feeder?
a) To distribute power to individual circuits in a building
b) To provide power to a specific piece of equipment
c) To connect the building's electrical system to the utility's power grid
d) To regulate voltage levels in a building's electrical system

Answer: a) To distribute power to individual circuits in a building
Explanation: An electrical feeder is a circuit that distributes power to multiple circuits within a building or facility.

Which of the following is a common type of electrical feeder?
a) Service entrance feeder
b) Branch circuit feeder
c) Panelboard feeder
d) Generator feeder

Answer: c) Panelboard feeder
Explanation: A panelboard feeder is a type of electrical feeder that connects a panelboard to a larger electrical system.

What is the maximum length of an electrical feeder, according to the National Electrical Code?
a) 50 feet
b) 100 feet
c) 150 feet
d) There is no maximum length specified

Answer: d) There is no maximum length specified
Explanation: The NEC does not specify a maximum length for electrical feeders, but it does have requirements for the size of the feeder based on the load it will be carrying.

What is the minimum size of an electrical feeder for a 200-amp panelboard, according to the NEC?

a) 2/0 AWG
b) 3/0 AWG
c) 4/0 AWG
d) 250 kcmil

Answer: b) 3/0 AWG
Explanation: According to NEC Table 310.15(B)(16), a 3/0 AWG copper conductor is the minimum size allowed for a 200-amp feeder.

Which of the following is a factor to consider when selecting the size of an electrical feeder?
a) The type of electrical load being served
b) The location of the panelboard being served
c) The number of circuit breakers in the panelboard
d) The color of the insulation on the feeder conductors

Answer: a) The type of electrical load being served
Explanation: The type of load being served (such as resistive, inductive, or capacitive) will affect the size of the feeder required to handle the load.

What is the minimum size copper conductor required for a 200-ampere feeder circuit that supplies continuous loads?
a. #2/0 AWG
b. #3/0 AWG
c. #4/0 AWG
d. #250 kcmil

Answer: c
Explanation: According to NEC Table 310.16, a 200-ampere continuous load requires a minimum size copper conductor of #4/0 AWG.

What is the maximum overcurrent protection device (OCPD) rating allowed for a 3/0 copper conductor feeder circuit supplying a 125-ampere continuous load?
a. 150 amperes
b. 175 amperes

c. 200 amperes

d. 225 amperes

Answer: b

Explanation: According to NEC 310.15(B)(16), a 3/0 copper conductor has an ampacity of 150 amperes, and the OCPD should not exceed 125% of this value for continuous loads, which is 187.5 amperes. However, the next standard OCPD size is 175 amperes, so this is the maximum allowed.

A feeder circuit that supplies a motor requires an ampacity of 175 amperes. What is the minimum size aluminum conductor required?

a. #2/0 AWG

b. #3/0 AWG

c. #4/0 AWG

d. #250 kcmil

Answer: d

Explanation: According to NEC Table 310.16, a 175-ampere load requires a minimum size aluminum conductor of #250 kcmil.

What is the maximum distance for a 3-phase, 480-volt feeder circuit with a 50-ampere load and a voltage drop of 3%?

a. 200 feet

b. 250 feet

c. 300 feet

d. 350 feet

Answer: c

Explanation: According to NEC Table 310.15(B)(16), a #8 AWG copper conductor has an ampacity of 50 amperes, and at 480 volts, the maximum distance for a 3% voltage drop is 300 feet.

What is the minimum size copper conductor required for a 125-ampere, 3-phase feeder circuit with a voltage drop of 2% and a distance of 200 feet?

a. #2 AWG

b. #3 AWG

c. #2/0 AWG

d. #3/0 AWG

Answer: c

Explanation: According to NEC Table 310.16, a 125-ampere load requires a minimum size copper conductor of #2/0 AWG. However, at a distance of 200 feet and a 2% voltage drop, the minimum size is increased to #2 AWG.

What is the maximum allowable voltage drop for branch circuits in the NEC?

a) 2%

b) 3%

c) 5%

d) 8%

Answer: b) 3%

Explanation: According to NEC 210.19(A)(1), the maximum voltage drop on branch circuits must not exceed 3% of the voltage supply.

What is the formula to calculate voltage drop?

a) $V = IR$

b) $V = IZ$

c) $V = IR + IZ$

d) $V = I(R + Z)$

Answer: d) $V = I(R + Z)$

Explanation: The formula to calculate voltage drop is $V = I(R + Z)$, where V is voltage, I is current, R is resistance, and Z is impedance.

What is the recommended maximum voltage drop for feeder circuits in the NEC?

a) 2%

b) 3%

c) 5%

d) 8%

Answer: c) 5%
Explanation: According to NEC 215.2(A)(4), the maximum voltage drop on feeder circuits must not exceed 5% of the voltage supply.

Which of the following factors can affect voltage drop?
a) Length of conductor
b) Conductor material
c) Temperature
d) All of the above

Answer: d) All of the above
Explanation: Voltage drop can be affected by several factors, including the length of the conductor, the conductor material, and temperature.

What is the typical voltage drop for a 100-foot length of 14 AWG copper wire carrying a load of 10 amps?
a) 0.50 volts
b) 1.0 volts
c) 1.5 volts
d) 2.0 volts

Answer: c) 1.5 volts
Explanation: Using the voltage drop formula V = I(R + Z), with R and Z values for 14 AWG copper wire and a 100-foot length, the voltage drop would be approximately 1.5 volts with a load of 10 amps.

What is the function of a circuit breaker in an electrical circuit?
a) to regulate the voltage
b) to regulate the current
c) to protect against overcurrent
d) to protect against overvoltage

Answer: c) to protect against overcurrent

Explanation: Circuit breakers are protective devices designed to protect electrical circuits from overcurrent, which can cause damage to the equipment or even start a fire. When a current overload occurs, the circuit breaker trips and interrupts the flow of electricity, preventing damage to the circuit.

Which of the following is a type of overcurrent protection device?
a) switch
b) fuse
c) relay
d) contactor

Answer: b) fuse
Explanation: Fuses are one of the most common types of overcurrent protection devices used in electrical circuits. They work by interrupting the flow of electricity when the current exceeds a certain level, thus protecting the circuit from damage.

What is the purpose of a ground fault circuit interrupter (GFCI)?
a) to prevent overvoltage
b) to prevent overcurrent
c) to prevent short circuits
d) to protect against electric shock

Answer: d) to protect against electric shock
Explanation: GFCIs are protective devices designed to protect people from electric shock. They work by quickly interrupting the flow of electricity if a ground fault is detected, which can occur if an electrical appliance or device comes into contact with water or if there is a fault in the wiring.

Which of the following is a type of overcurrent?
a) overvoltage
b) short circuit
c) undervoltage
d) open circuit

Answer: b) short circuit

Explanation: A short circuit is an overcurrent condition that occurs when an electrical circuit is accidentally or intentionally completed through a path of low resistance, bypassing the load. This can cause an excessive flow of current that can damage the circuit or even start a fire.

What is the difference between a fuse and a circuit breaker?
a) fuses are more reliable than circuit breakers
b) circuit breakers are easier to reset than fuses
c) fuses provide more precise protection than circuit breakers
d) circuit breakers are reusable, while fuses are not

Answer: d) circuit breakers are reusable, while fuses are not

Explanation: Circuit breakers are reusable overcurrent protection devices that can be reset after they have tripped. Fuses, on the other hand, are one-time-use devices that must be replaced once they have blown. While circuit breakers are generally more convenient and easier to use than fuses, they may not provide as precise protection for sensitive equipment.

What is the main purpose of grounding in electrical systems?
a) To provide a path for current to flow in the event of a fault
b) To provide a neutral point for the electrical system
c) To regulate voltage levels
d) To protect equipment from power surges

Answer: a) To provide a path for current to flow in the event of a fault

Explanation: Grounding provides a low-impedance path for current to flow in the event of a fault, which helps to protect people and equipment from electrical hazards.

Which of the following is an example of an equipment grounding conductor?
a) A neutral wire
b) A bare copper wire
c) A hot wire
d) A white wire

Answer: b) A bare copper wire

Explanation: An equipment grounding conductor is a wire that provides a low-impedance path to ground for electrical equipment. It is typically a bare copper wire, although it may be green or green/yellow striped.

What is the maximum resistance allowed for a grounding electrode?

a) 1 ohm
b) 5 ohms
c) 10 ohms
d) 25 ohms

Answer: b) 5 ohms

Explanation: The NEC recommends that the resistance to ground of a grounding electrode not exceed 25 ohms. However, for sensitive electronic equipment, a lower resistance of 5 ohms or less is recommended.

Which of the following is an example of a grounding electrode?

a) A copper water pipe
b) A PVC conduit
c) A flexible metal conduit
d) A non-metallic sheathed cable

Answer: a) A copper water pipe

Explanation: A grounding electrode is a component of the grounding system that provides a connection to earth. Common examples include metal water pipes, metal rods, and metal plates.

What is the difference between a ground fault and a short circuit?

a) A ground fault occurs when current flows to ground, while a short circuit occurs when current flows through a low-impedance path.
b) A ground fault occurs when current flows through a low-impedance path, while a short circuit occurs when current flows to ground.
c) A ground fault and a short circuit are the same thing.
d) A ground fault and a short circuit are both caused by improper grounding.

Answer: a) A ground fault occurs when current flows to ground, while a short circuit occurs when current flows through a low-impedance path.

Explanation: A ground fault occurs when current flows from an electrical circuit to ground. This can happen when there is a break in the insulation or other fault in the circuit. A short circuit, on the other hand, occurs when current flows through a low-impedance path, such as a direct connection between a hot wire and a neutral wire.

What type of protective device is used to limit overvoltage conditions and protect equipment from electrical surges?
a) Circuit breaker
b) Fuse
c) Surge protector
d) Ground fault circuit interrupter

Answer: c) Surge protector

Explanation: A surge protector is a device that limits overvoltage conditions, protects equipment from electrical surges and prevents current from flowing to the ground.

What type of protective device is designed to open a circuit if the current exceeds a certain level?
a) Circuit breaker
b) Fuse
c) Surge protector
d) Ground fault circuit interrupter

Answer: a) Circuit breaker

Explanation: A circuit breaker is a protective device that is designed to open a circuit if the current exceeds a certain level, thereby preventing electrical fires and damage to equipment.

What type of protective device is a safety switch that automatically disconnects a circuit when it detects an imbalance in the electrical current?
a) Circuit breaker
b) Fuse

c) Surge protector
d) Ground fault circuit interrupter

Answer: d) Ground fault circuit interrupter
Explanation: A ground fault circuit interrupter (GFCI) is a safety switch that is designed to automatically disconnect a circuit when it detects an imbalance in the electrical current, thereby preventing electrical shocks and electrocutions.

What type of protective device is a device that melts and opens a circuit when the current exceeds a certain level?
a) Circuit breaker
b) Fuse
c) Surge protector
d) Ground fault circuit interrupter

Answer: b) Fuse
Explanation: A fuse is a protective device that melts and opens a circuit when the current exceeds a certain level, thereby preventing electrical fires and damage to equipment.

What type of protective device is used to protect motors and other electrical equipment from damage caused by overvoltage, undervoltage, or voltage imbalances?
a) Circuit breaker
b) Fuse
c) Surge protector
d) Voltage relay

Answer: d) Voltage relay
Explanation: A voltage relay is a protective device that is used to protect motors and other electrical equipment from damage caused by overvoltage, undervoltage, or voltage imbalances. It works by monitoring the voltage levels in the electrical system and triggering an alarm or disconnecting the equipment if the voltage levels are outside of acceptable parameters.

What is the purpose of overcurrent protection devices?
A) To protect against short circuits
B) To protect against ground faults
C) To protect against overloads

D) All of the above

What type of overcurrent protection device is typically used to protect a motor?
A) Fuses
B) Circuit breakers
C) Ground fault circuit interrupters (GFCIs)
D) Surge protectors

What is the difference between a circuit breaker and a fuse?
A) A circuit breaker can be reset, while a fuse must be replaced.
B) A circuit breaker is more expensive than a fuse.
C) A circuit breaker can only be used for low voltage applications, while a fuse can be used for both low and high voltage.
D) A circuit breaker is more sensitive than a fuse.

Which of the following is NOT a factor to consider when selecting an overcurrent protection device?
A) The size of the conductor being protected
B) The type of load being protected
C) The ambient temperature where the device will be installed
D) The color of the device

Answer: D

Explanation: The color of the device is not a factor to consider when selecting an overcurrent protection device. Factors such as conductor size, load type, and ambient temperature are important to ensure the device provides proper protection.

What is the purpose of a ground fault circuit interrupter (GFCI)?

A) To protect against short circuits
B) To protect against ground faults
C) To protect against overloads
D) All of the above

Answer: B

Explanation: A GFCI is a type of overcurrent protection device that is designed to protect against ground faults, which occur when an electrical current flows through a person or object to ground.

Which of the following is the correct formula for calculating the minimum size of a branch circuit conductor?

a) (Load in VA) / (Voltage)
b) (Load in VA) / (Voltage x 1.15)
c) (Load in watts) / (Voltage)
d) (Load in watts) / (Voltage x 1.15)

Answer: d

Explanation: The minimum size of a branch circuit conductor can be calculated using the formula (Load in watts) / (Voltage x 1.15) as per the NEC.

When sizing branch circuit conductors, which of the following factors should be taken into account?

a) Voltage drop and current carrying capacity
b) Voltage drop and power factor
c) Current carrying capacity and impedance
d) Power factor and impedance

Answer: a

Explanation: When sizing branch circuit conductors, voltage drop and current carrying capacity should be taken into account.

What is the minimum size of a branch circuit conductor for a 20A lighting load in a dwelling unit?
a) 12 AWG
b) 10 AWG
c) 8 AWG
d) 6 AWG

Answer: a

Explanation: According to the NEC, a 20A lighting load in a dwelling unit requires a minimum branch circuit conductor size of 12 AWG.

Which of the following factors can increase the required ampacity of a branch circuit conductor?
a) Higher ambient temperature
b) Lower ambient temperature
c) Longer circuit length
d) Lower circuit voltage

Answer: a

Explanation: Higher ambient temperature can increase the required ampacity of a branch circuit conductor, as per the NEC.

What is the minimum size of a branch circuit conductor for a 30A, 240V load in a non-dwelling unit?
a) 10 AWG
b) 8 AWG
c) 6 AWG
d) 4 AWG

Answer: b

Explanation: According to the NEC, a 30A, 240V load in a non-dwelling unit requires a minimum branch circuit conductor size of 8 AWG.

What is the most commonly used type of conductor in electrical wiring?
a) Aluminum
b) Copper
c) Silver
d) Gold

Answer: b) Copper
Explanation: Copper is the most commonly used conductor in electrical wiring due to its excellent conductivity and durability.

Which of the following types of conductors is often used in high temperature applications?
a) Aluminum
b) Copper
c) Silver
d) Nickel

Answer: d) Nickel
Explanation: Nickel is often used in high temperature applications due to its ability to resist oxidation and maintain its strength at high temperatures.

What is the primary function of a ground conductor in an electrical circuit?
a) To carry current under normal operating conditions
b) To provide overcurrent protection
c) To complete the circuit and return current to its source
d) To provide a safe path for fault currents to travel

Answer: d) To provide a safe path for fault currents to travel
Explanation: The ground conductor in an electrical circuit is designed to provide a safe path for fault currents to travel, which can help prevent electrical shocks and other hazards.

What is the purpose of a stranded conductor in electrical wiring?
a) To increase the conductivity of the wire
b) To reduce the overall cost of the wiring installation

c) To make the wire more flexible and easier to install
d) To provide additional insulation for the wire

Answer: c) To make the wire more flexible and easier to install
Explanation: Stranded conductors are made up of multiple smaller wires twisted together, which makes the wire more flexible and easier to install than a solid conductor of the same size.

What is the maximum temperature rating for a typical PVC insulated wire used in electrical wiring?
a) 60°C
b) 75°C
c) 90°C
d) 105°C

Answer: c) 90°C
Explanation: PVC insulated wire is typically rated for a maximum temperature of 90°C, which is suitable for most residential and commercial wiring applications.

Which type of circuit breaker is typically used for low voltage applications?
a. Air circuit breaker
b. Magnetic circuit breaker
c. Thermal circuit breaker
d. Miniature circuit breaker

Which of the following is a type of switch that is commonly used in residential wiring for controlling light fixtures and outlets?
a) Single-pole switch
b) Double-pole switch
c) Three-way switch
d) Four-way switch

Answer: a) Single-pole switch

Explanation: Single-pole switches are the most commonly used type of switch in residential wiring for controlling light fixtures and outlets. They have two terminals and are used for simple on/off control of a circuit.

Which of the following is true about a double-pole switch?
a) It has two separate circuits that can be controlled independently.
b) It has four terminals and is used to control a circuit from multiple locations.
c) It is used for simple on/off control of a circuit.
d) It is not commonly used in residential wiring.

Answer: a) It has two separate circuits that can be controlled independently.

Explanation: A double-pole switch has two separate circuits that can be controlled independently. It is commonly used in applications such as controlling a 240-volt circuit or a circuit with a high current load.

Which of the following is a type of switch that is used for controlling a circuit from multiple locations?
a) Single-pole switch
b) Double-pole switch
c) Three-way switch
d) Four-way switch

Answer: c) Three-way switch

Explanation: A three-way switch is used for controlling a circuit from two different locations. It has three terminals and is often used in stairways, hallways, and large rooms.

Which of the following is true about a four-way switch?
a) It has two terminals and is used for simple on/off control of a circuit.
b) It has four terminals and is used to control a circuit from multiple locations.
c) It is used for controlling a 240-volt circuit.
d) It is not commonly used in residential wiring.

Answer: b) It has four terminals and is used to control a circuit from multiple locations.

Explanation: A four-way switch is used for controlling a circuit from three or more locations. It has four terminals and is often used in larger rooms and hallways.

Which of the following types of switches is commonly used for controlling ceiling fans?
a) Single-pole switch
b) Double-pole switch
c) Three-way switch
d) Fan speed control switch

Answer: d) Fan speed control switch

Explanation: Fan speed control switches are specifically designed for controlling the speed of a ceiling fan. They often have multiple settings and may include a separate switch for controlling the fan light.Answer: d
Explanation: Miniature circuit breakers are typically used for low voltage applications such as in homes and offices.

What type of electrical device is used to protect electrical circuits from overvoltage?
a. Fuse
b. Capacitor
c. Surge protector
d. Diode

Answer: c
Explanation: Surge protectors are designed to protect electrical circuits and equipment from overvoltage, which can occur during power surges.

Which type of electrical device is used to convert DC voltage to AC voltage?
a. Transformer
b. Rectifier

c. Inverter

d. Capacitor

Answer: c

Explanation: Inverters are used to convert DC voltage to AC voltage. They are commonly used in solar power systems and other applications where DC power needs to be converted to AC power.

Which electrical device is used to measure electrical current?

a. Voltmeter

b. Ammeter

c. Ohmmeter

d. Multimeter

Answer: b

Explanation: An ammeter is used to measure the amount of electrical current flowing in a circuit.

What type of electrical device is used to control the speed of an electric motor?

a. Relay

b. Circuit breaker

c. Capacitor

d. Variable frequency drive

Answer: d

Explanation: A variable frequency drive (VFD) is used to control the speed of an electric motor by adjusting the frequency of the electrical supply to the motor.

Which of the following is NOT a type of fuse?

a) Cartridge fuse

b) Thermal fuse

c) Resettable fuse

d) Capacitor fuse

Answer: d) Capacitor fuse

Explanation: Cartridge fuses, thermal fuses, and resettable fuses are all common types of fuses used in electrical systems. Capacitor fuses are not a type of fuse.

What is the purpose of a fuse in an electrical circuit?
a) To regulate voltage
b) To regulate current
c) To provide an electrical connection
d) To protect against overcurrent

Answer: d) To protect against overcurrent

Explanation: Fuses are designed to break the circuit in the event of overcurrent, which could potentially cause damage to the electrical equipment or create a safety hazard.

Which of the following is NOT a factor in selecting a fuse for a particular application?
a) Voltage rating
b) Current rating
c) Type of load
d) Color of the fuse

Answer: d) Color of the fuse

Explanation: Voltage rating, current rating, and type of load are all important factors in selecting a fuse for a particular application. The color of the fuse, on the other hand, is usually not a significant factor.

Which type of fuse is commonly used in automotive applications?
a) Cartridge fuse
b) Thermal fuse
c) Resettable fuse

d) Blade fuse

Answer: d) Blade fuse

Explanation: Blade fuses are commonly used in automotive applications, where they are often found in fuse boxes and other electrical components.

Which of the following is an advantage of resettable fuses over traditional fuses?
a) They are smaller in size
b) They are more reliable
c) They do not need to be replaced after a fault
d) They are more precise in their response to overcurrent

Answer: c) They do not need to be replaced after a fault

Explanation: Resettable fuses are designed to "reset" themselves after a fault, rather than requiring replacement like traditional fuses. This can save time and money in the long run, although resettable fuses may not be appropriate for all applications.

Which of the following is NOT a common type of circuit breaker?
a) Magnetic
b) Thermal
c) Hydraulic
d) Hybrid

Answer: c) Hydraulic

Explanation: Magnetic and thermal circuit breakers are common types of overcurrent protection devices. Magnetic circuit breakers operate based on the strength of the magnetic field created by the current, while thermal circuit breakers operate based on the heat generated by the current. Hybrid circuit breakers combine the features of both magnetic and thermal breakers. Hydraulic circuit breakers, on the other hand, use a hydraulic fluid to trip the breaker in response to an overcurrent condition.

What is the purpose of a circuit breaker?
a) To regulate the voltage in a circuit
b) To prevent overloads and short circuits
c) To increase the current-carrying capacity of a circuit
d) To reduce the resistance of a circuit

Answer: b) To prevent overloads and short circuits

Explanation: A circuit breaker is an overcurrent protection device that is designed to automatically open and interrupt the flow of current in a circuit when an overload or short circuit occurs. This helps to prevent damage to the circuit and the devices connected to it.

Which of the following is a common type of trip mechanism for a circuit breaker?
a) Thermal
b) Magnetic
c) Thermal-magnetic
d) All of the above

Answer: d) All of the above

Explanation: Circuit breakers can use a variety of trip mechanisms, including thermal, magnetic, and thermal-magnetic. Thermal breakers operate based on the heat generated by the current, magnetic breakers operate based on the strength of the magnetic field created by the current, and thermal-magnetic breakers use a combination of both mechanisms.

Which of the following is NOT a factor in selecting a circuit breaker for a particular application?
a) Voltage rating
b) Current rating
c) Phase
d) Temperature rating

Answer: d) Temperature rating

Explanation: When selecting a circuit breaker for a particular application, it is important to consider the voltage and current rating of the circuit, as well as the phase (single or three-phase). The temperature rating is not typically a factor in selecting a circuit breaker, although it is important to ensure that the breaker is rated for the expected ambient temperature.

What is the difference between a single-pole and a double-pole circuit breaker?
a) The number of wires that can be connected to the breaker
b) The voltage rating of the breaker
c) The current rating of the breaker
d) The number of circuits that can be protected by the breaker

Answer: d) The number of circuits that can be protected by the breaker

Explanation: A single-pole circuit breaker can protect a single circuit, while a double-pole circuit breaker can protect two circuits. The number of wires that can be connected to the breaker, as well as the voltage and current ratings, may vary depending on the specific type of breaker.

What is the main purpose of a transformer?
a) To generate electricity
b) To reduce or increase voltage
c) To protect electrical equipment
d) To control power factor

Answer: b
Explanation: The main purpose of a transformer is to reduce or increase voltage in an electrical circuit.

Which type of transformer is used to convert high voltage to low voltage?
a) Step-up transformer
b) Step-down transformer
c) Autotransformer
d) Distribution transformer

Answer: b

Explanation: A step-down transformer is used to convert high voltage to low voltage.

Which of the following is not a common type of transformer core?
a) Air
b) Iron
c) Ferrite
d) Copper

Answer: d

Explanation: Copper is not a common type of transformer core. Air, iron, and ferrite are commonly used.

What is the primary function of a transformer's core?
a) To reduce eddy current losses
b) To increase transformer efficiency
c) To reduce the weight of the transformer
d) To reduce the magnetic flux

Answer: b

Explanation: The primary function of a transformer's core is to increase transformer efficiency.

What is the difference between a single-phase transformer and a three-phase transformer?
a) A single-phase transformer has one primary and one secondary winding, while a three-phase transformer has three primary and three secondary windings.
b) A single-phase transformer has one primary and three secondary windings, while a three-phase transformer has three primary and one secondary winding.
c) A single-phase transformer has one primary and one secondary winding, while a three-phase transformer has three primary and one secondary winding.
d) A single-phase transformer has one primary and three secondary windings, while a three-phase transformer has three primary and three secondary windings.

Answer: a

Explanation: A single-phase transformer has one primary and one secondary winding, while a three-phase transformer has three primary and three secondary windings.

What is the purpose of a capacitor start motor?
a. To provide constant torque at high speeds
b. To provide high starting torque
c. To provide variable speed control
d. To provide high efficiency

Answer: b
Explanation: A capacitor start motor uses a capacitor to provide an initial high starting torque, which allows the motor to overcome inertia and start rotating.

What is the most common type of motor used in industrial applications?
a. Single-phase induction motor
b. Three-phase induction motor
c. Synchronous motor
d. DC motor

Answer: b
Explanation: Three-phase induction motors are the most common type of motor used in industrial applications because they are reliable, efficient, and relatively inexpensive.

What is the purpose of a motor overload protection device?
a. To protect the motor from overvoltage
b. To protect the motor from undervoltage
c. To protect the motor from overcurrent
d. To protect the motor from overheating

Answer: d
Explanation: A motor overload protection device is designed to protect the motor from overheating due to excessive current, which can be caused by a variety of factors such as overloading or a mechanical problem.

What is the difference between a squirrel-cage motor and a wound-rotor motor?

a. A squirrel-cage motor has a wound rotor, while a wound-rotor motor has a squirrel-cage rotor.

b. A squirrel-cage motor has a stationary rotor, while a wound-rotor motor has a rotating rotor.

c. A squirrel-cage motor has a simple, rugged design, while a wound-rotor motor has a more complex design.

d. A squirrel-cage motor is less efficient than a wound-rotor motor.

Answer: c

Explanation: A squirrel-cage motor has a simple, rugged design with a stationary rotor that does not require brushes or slip rings, while a wound-rotor motor has a more complex design with a rotating rotor that includes brushes and slip rings.

What is the purpose of a VFD (variable frequency drive) in a motor control system?

a. To provide overload protection

b. To provide soft-starting and speed control

c. To provide phase conversion

d. To provide surge protection

Answer: b

Explanation: A VFD (variable frequency drive) is used in motor control systems to provide soft-starting and speed control by varying the frequency of the electrical power supplied to the motor. This helps to improve energy efficiency and reduce wear and tear on the motor.

Which of the following is not a type of generator?

a. AC generator

b. DC generator

c. Synchronous generator

d. Ground-fault generator

Answer: d

Explanation: A ground-fault generator is not a type of generator. AC, DC, and synchronous generators are commonly used in various electrical applications.

What is the primary source of energy for a generator?
a. Chemical energy
b. Electrical energy
c. Mechanical energy
d. Thermal energy

Answer: c

Explanation: The primary source of energy for a generator is mechanical energy, usually from an engine or turbine.

What is the function of the AVR in a generator?
a. To regulate the voltage output of the generator
b. To start and stop the generator automatically
c. To provide overload protection for the generator
d. To provide short-circuit protection for the generator

Answer: a

Explanation: The AVR (automatic voltage regulator) in a generator is responsible for regulating the voltage output of the generator, to ensure that it remains within safe and appropriate limits.

Which of the following factors affects the frequency of a generator?
a. The number of poles in the generator
b. The resistance of the generator
c. The voltage of the generator
d. The current output of the generator

Answer: a

Explanation: The frequency of a generator is determined by the number of poles in the generator and the speed at which the generator is rotating.

Which of the following is a common fuel used to power generators?
a. Propane
b. Diesel
c. Gasoline
d. All of the above

Answer: d

Explanation: Propane, diesel, and gasoline are all commonly used fuels for generators, depending on the application and availability.

What is the minimum size of wire required for a 20-ampere circuit protected by a fuse or circuit breaker?
a. #14 AWG
b. #12 AWG
c. #10 AWG
d. #8 AWG

Answer: b. #12 AWG
Explanation: According to the National Electrical Code (NEC), a 20-ampere circuit must be protected by a fuse or circuit breaker and must have a minimum wire size of #12 AWG.

What is the maximum allowable distance between junction boxes in a conduit run?
a. 50 feet
b. 75 feet
c. 100 feet
d. 150 feet

Answer: a. 50 feet

Explanation: The NEC specifies that junction boxes must be installed at intervals not exceeding 100 feet for straight conduit runs and 50 feet for conduit runs with bends or turns.

Which type of conduit is most commonly used for underground wiring?
a. Rigid metal conduit (RMC)
b. Electrical metallic tubing (EMT)
c. Intermediate metal conduit (IMC)
d. PVC conduit

Answer: d. PVC conduit
Explanation: PVC conduit is commonly used for underground wiring due to its resistance to corrosion and moisture.

What is the maximum allowable unsupported span of #10 AWG copper wire in a residential attic?
a. 5 feet
b. 7 feet
c. 9 feet
d. 11 feet

Answer: b. 7 feet
Explanation: According to the NEC, #10 AWG copper wire must be supported every 7.5 feet when run in a residential attic.

What is the minimum bend radius for flexible metal conduit (FMC)?
a. 3 inches
b. 4 inches
c. 5 inches
d. 6 inches

Answer: b. 4 inches
Explanation: The NEC specifies that flexible metal conduit (FMC) must have a minimum bend radius of 4 inches.

Which of the following is a common type of electric motor used in industrial applications?
a) Shaded pole motor
b) Single-phase motor
c) Three-phase motor
d) Universal motor

Answer: c) Three-phase motor
Explanation: Three-phase motors are commonly used in industrial applications due to their high efficiency, low maintenance, and ability to operate at constant speed.

Which type of switch is commonly used to control lighting circuits in residential buildings?
a) Toggle switch
b) Push-button switch
c) Rotary switch
d) Dimmer switch

Answer: a) Toggle switch
Explanation: Toggle switches are a common type of switch used to control lighting circuits in residential buildings. They are simple to use and typically have a longer lifespan than other types of switches.

What is the primary function of a ground fault circuit interrupter (GFCI)?
a) To protect against overcurrent
b) To protect against short circuits
c) To protect against ground faults
d) To protect against electrical arcing

Answer: c) To protect against ground faults
Explanation: GFCIs are designed to protect against ground faults, which occur when electrical current leaks to ground through an unintended path, such as a person. They are commonly used in areas where water is present, such as kitchens, bathrooms, and outdoor locations.

What is the maximum voltage rating for a standard household electrical outlet in North America?
a) 110 volts
b) 120 volts
c) 220 volts
d) 240 volts

Answer: b) 120 volts
Explanation: The standard voltage rating for a household electrical outlet in North America is 120 volts. This is also known as the nominal voltage, as the actual voltage may vary slightly due to factors such as electrical load and distance from the electrical panel.

Which of the following materials is commonly used for electrical wiring in residential and commercial buildings?
a) Copper
b) Aluminum
c) Steel
d) PVC

Answer: a) Copper
Explanation: Copper is a common material used for electrical wiring due to its high conductivity, durability, and resistance to corrosion. Aluminum wiring is also used in some applications, but it is less common due to its lower conductivity and increased risk of oxidation.

What is the difference between a 3-phase 208V system and a 1-phase 208V system?
a. The 3-phase system has a higher voltage than the 1-phase system
b. The 1-phase system has a higher voltage than the 3-phase system
c. The 3-phase system has three conductors and the 1-phase system has two conductors
d. The 1-phase system has three conductors and the 3-phase system has two conductors

Answer: c

Explanation: A 3-phase 208V system has three conductors (L1, L2, and L3) and a voltage of 208V between any two conductors, while a 1-phase 208V system has two conductors (L1 and L2) and a voltage of 208V between them.

What is the maximum amperage for a 12 AWG copper conductor at 60°C?
a. 15A
b. 20A
c. 25A
d. 30A

Answer: b

Explanation: According to the National Electric Code (NEC), the maximum amperage for a 12 AWG copper conductor at 60°C is 20A.

Which of the following electrical equipment is rated in horsepower (HP)?
a. Motors
b. Circuit breakers
c. Fuses
d. Switches

Answer: a

Explanation: Motors are rated in horsepower (HP) to indicate their power output.

What is the difference between a type MC cable and a type AC cable?
a. Type MC cable has a metal sheath, while type AC cable has a non-metallic sheath
b. Type AC cable has a metal sheath, while type MC cable has a non-metallic sheath
c. Type MC cable is used for high voltage applications, while type AC cable is used for low voltage applications
d. Type AC cable is used for high voltage applications, while type MC cable is used for low voltage applications

Answer: a

Explanation: Type MC cable has a metal sheath and is used for a variety of applications, including branch circuits, feeders, and services. Type AC cable, on the other hand, has a flexible metallic armor and is typically used for branch circuits and feeders in residential and commercial applications.

What is the maximum number of current-carrying conductors allowed in a raceway or cable tray?
a. 1
b. 2
c. 3
d. As many as can fit

Answer: d

Explanation: There is no maximum number of current-carrying conductors allowed in a raceway or cable tray, as long as the conductors are not overloaded and the temperature rise is within the acceptable limits specified in the National Electric Code (NEC).

What is the purpose of a GFCI?
a) To regulate voltage in an electrical circuit
b) To prevent overloading of electrical circuits
c) To protect against electrical shock
d) To increase the efficiency of electrical systems

Answer: c) To protect against electrical shock

Explanation: GFCI (ground fault circuit interrupter) is a safety device that is designed to protect against electrical shock. It works by monitoring the current flowing through a circuit and tripping the circuit if it detects a current imbalance, such as when current flows through a person instead of the intended path.

What is the maximum amperage rating for a standard residential GFCI receptacle?
a) 15 amps

b) 20 amps

c) 30 amps

d) 50 amps

Answer: b) 20 amps

Explanation: The maximum amperage rating for a standard residential GFCI receptacle is 20 amps. This is because GFCIs are typically used to protect small appliances and other household devices, which generally do not require more than 20 amps of power.

What is the difference between a GFCI receptacle and a GFCI circuit breaker?

a) A GFCI receptacle is a type of outlet, while a GFCI circuit breaker is a type of switch

b) A GFCI receptacle provides protection for all devices connected to it, while a GFCI circuit breaker provides protection for an entire circuit

c) A GFCI receptacle can only be used in new construction, while a GFCI circuit breaker can be used in both new and existing construction

d) A GFCI receptacle is more expensive than a GFCI circuit breaker

Answer: b) A GFCI receptacle provides protection for all devices connected to it, while a GFCI circuit breaker provides protection for an entire circuit

Explanation: A GFCI receptacle is an outlet with built-in GFCI protection, which provides protection for all devices connected to it. A GFCI circuit breaker, on the other hand, provides protection for an entire circuit. Both types of protection can be used together for added safety.

Which of the following types of electrical equipment typically requires GFCI protection?

a) Outdoor receptacles

b) Refrigerators

c) Central air conditioning units

d) Electric water heaters

Answer: a) Outdoor receptacles

Explanation: GFCI protection is typically required for outdoor receptacles, as they are more susceptible to exposure to moisture and other environmental hazards. This added protection helps prevent electrical shock in wet or damp conditions.

Which of the following statements is true about testing a GFCI receptacle?
a) GFCI receptacles should be tested every five years
b) Testing a GFCI receptacle involves pressing the "test" button and checking that the power is disconnected
c) Testing a GFCI receptacle involves pressing the "reset" button and checking that the power is restored
d) If a GFCI receptacle fails a test, it must be replaced immediately

Answer: b) Testing a GFCI receptacle involves pressing the "test" button and checking that the power is disconnected

Explanation: To test a GFCI receptacle, simply press the "test" button and check that power is disconnected to the device plugged in to the receptacle. If power is still available, the GFCI receptacle may be faulty and should be replaced. GFCI receptacles should be tested regularly to ensure that they are providing adequate protection against electrical shock.

Which of the following wiring methods is permitted for use in a hazardous location?
a. Non-metallic sheathed cable
b. Rigid metal conduit
c. Liquidtight flexible metal conduit
d. Flexible metallic tubing

Answer: c. Liquidtight flexible metal conduit

Explanation: Hazardous locations require specific wiring methods to ensure safety. Liquidtight flexible metal conduit is permitted for use in these locations due to its ability to provide protection against moisture and other hazards.

What is the maximum number of conductors that can be installed in a single conduit using the Table 1 allowable fill percentages for raceways?
a. 4

b. 6

c. 8

d. 10

Answer: b. 6

Explanation: Table 1 in the National Electrical Code provides allowable fill percentages for raceways. The maximum number of conductors that can be installed in a single conduit is determined by these fill percentages, and in most cases, it is limited to 40% fill. Using the Table 1 allowable fill percentages, the maximum number of conductors that can be installed in a single conduit is 6.

Which of the following wiring methods is required to be protected by a metal or nonmetallic conduit?

a. Type NM cable

b. Type AC cable

c. Type MC cable

d. Type UF cable

Answer: c. Type MC cable

Explanation: Type MC cable must be protected by a metal or nonmetallic conduit due to its construction. Type NM, Type AC, and Type UF cables do not require conduit protection.

Which of the following wiring methods is permitted for use in a damp location?

a. Liquidtight flexible nonmetallic conduit

b. Rigid metal conduit

c. Intermediate metal conduit

d. Flexible metallic tubing

Answer: a. Liquidtight flexible nonmetallic conduit

Explanation: Damp locations require specific wiring methods to ensure safety. Liquidtight flexible nonmetallic conduit is permitted for use in these locations due to its ability to provide protection against moisture and dampness.

Which of the following wiring methods is typically used for exposed work?
a. Non-metallic sheathed cable
b. Rigid metal conduit
c. Liquidtight flexible metal conduit
d. Flexible metallic tubing

Answer: b. Rigid metal conduit

Explanation: Rigid metal conduit is typically used for exposed work due to its durability and ability to provide protection against physical damage. Non-metallic sheathed cable, liquidtight flexible metal conduit, and flexible metallic tubing are often used for concealed work.

Which of the following factors affect motor efficiency?
a) Motor design and construction
b) Operating conditions
c) Maintenance practices
d) All of the above

Answer: d) All of the above
Explanation: Motor efficiency is affected by a variety of factors, including the design and construction of the motor itself, the operating conditions under which it is used, and the maintenance practices that are performed on it over time.

Which type of motor generally has the highest efficiency?
a) AC induction motor
b) DC motor
c) Brushless DC motor
d) Stepper motor

Answer: c) Brushless DC motor

Explanation: Brushless DC motors are known for their high efficiency, due in part to their lack of brushes and commutators. They are often used in applications that require high levels of precision and reliability.

How is motor efficiency typically expressed?
a) In watts per hour
b) In joules per second
c) In horsepower
d) In a percentage

Answer: d) In a percentage
Explanation: Motor efficiency is typically expressed as a percentage of the input power that is converted into useful output power. For example, if a motor is 85% efficient, it is able to convert 85% of the electrical power it receives into mechanical power at the shaft.

What is the relationship between motor efficiency and energy consumption?
a) Higher efficiency motors consume more energy
b) Lower efficiency motors consume more energy
c) Motor efficiency has no impact on energy consumption
d) It depends on the specific application and usage patterns

Answer: b) Lower efficiency motors consume more energy
Explanation: Motors with lower efficiency ratings will consume more energy in order to produce the same amount of mechanical work. This is because a higher percentage of the input power is lost as heat, noise, and vibration, rather than being converted into useful work at the output shaft.

Which of the following measures can be taken to improve motor efficiency?
a) Upgrading to a more efficient motor design
b) Reducing the load on the motor
c) Using variable speed drives to match motor speed to load requirements
d) All of the above

Answer: d) All of the above

Explanation: Upgrading to a more efficient motor design, reducing the load on the motor, and using variable speed drives to match motor speed to load requirements are all effective measures for improving motor efficiency. These measures can also help to reduce energy consumption and extend the lifespan of the motor.

What is the purpose of motor protection?
a) To prevent overloading of the motor
b) To protect the motor against short circuits and overloads
c) To prevent power outages caused by the motor
d) To reduce the cost of energy used by the motor

Answer: b

Explanation: Motor protection devices are used to prevent damage to the motor caused by overloads, short circuits, ground faults, and other types of electrical faults. These devices include overload relays, fuses, and circuit breakers, which are designed to trip and disconnect the motor from the power source in the event of a fault.

Which of the following motor protection devices can be reset manually after tripping?
a) Thermal overload relay
b) Fuses
c) Circuit breaker
d) Ground fault relay

Answer: c

Explanation: Circuit breakers are motor protection devices that can be manually reset after tripping due to an overload or short circuit. This makes them useful for applications where it is necessary to quickly restart the motor after a fault has been corrected.

What is the primary function of a ground fault relay?
a) To protect the motor against overloads
b) To protect the motor against short circuits
c) To prevent electrical shock to personnel
d) To prevent the motor from overheating

Answer: c

Explanation: Ground fault relays are used to detect electrical faults that could result in electrical shock to personnel. These devices are designed to quickly disconnect the motor from the power source in the event of a ground fault.

Which of the following motor protection devices is typically used for low-voltage motors?
a) Fuses
b) Circuit breakers
c) Thermal overload relays
d) Ground fault relays

Answer: c

Explanation: Thermal overload relays are typically used for low-voltage motors to protect against overloads. These devices are designed to trip and disconnect the motor from the power source if the current drawn by the motor exceeds a certain level.

Which of the following motor protection devices is typically used for high-voltage motors?
a) Fuses
b) Circuit breakers
c) Thermal overload relays
d) Ground fault relays

Answer: b

Explanation: Circuit breakers are typically used for high-voltage motors to protect against overloads and short circuits. These devices are designed to trip and disconnect the motor from the power source if the current drawn by the motor exceeds a certain level or a fault occurs.

What is the primary purpose of a motor control circuit?
a) To protect the motor from overloading
b) To control the speed of the motor
c) To provide a safe means of starting and stopping the motor
d) To limit the amount of electrical energy used by the motor

Answer: c
Explanation: The primary purpose of a motor control circuit is to provide a safe means of starting and stopping the motor.

What is the function of a disconnect switch?
a) To control the speed of a motor
b) To protect against short circuits and overloads
c) To provide a means of disconnecting power to a circuit
d) To monitor the current flow in a circuit

Answer: c
Explanation: A disconnect switch provides a means of disconnecting power to a circuit for safety purposes, such as during maintenance or repairs.

Which of the following is not a common type of control device used in electrical systems?
a) Toggle switch
b) Relay
c) Circuit breaker
d) Timer

Answer: c
Explanation: Circuit breakers are not typically used as control devices, as their primary function is to protect against overloads and short circuits.

Which of the following is a type of overload protection commonly used in motor control circuits?
a) Fuses

b) Ground fault circuit interrupters (GFCIs)

c) Surge protectors

d) Capacitors

Answer: a

Explanation: Fuses are commonly used to provide overload protection in motor control circuits by breaking the circuit when the current exceeds a certain level.

What is the purpose of a push button control station?

a) To start and stop a motor

b) To monitor the current flow in a circuit

c) To provide a means of disconnecting power to a circuit

d) To control the speed of a motor

Answer: a

Explanation: A push button control station is used to start and stop a motor in a motor control circuit.

When installing electrical equipment and devices, which of the following should be considered to ensure safety and compliance with codes and regulations?

a) Proper grounding and bonding

b) Use of non-approved materials

c) Overloading circuits

d) Improperly sized conductors

Answer: a) Proper grounding and bonding

Explanation: Proper grounding and bonding are essential for safety and compliance with electrical codes and regulations. Grounding provides a safe path for electrical current to flow in case of a fault, while bonding connects all metal parts of a system to prevent the buildup of potentially dangerous electrical charges.

What is the primary purpose of conduit in an electrical installation?

a) To protect the conductors from damage

b) To increase the conductivity of the conductors

c) To decrease the cost of the installation

d) To provide additional insulation to the conductors

Answer: a) To protect the conductors from damage

Explanation: Conduit is used in electrical installations to protect the conductors from damage due to physical or environmental factors, such as moisture or rodents. Conduit can also be used to contain conductors and prevent them from interfering with other electrical systems.

When installing electrical equipment, what is the purpose of a lockout/tagout procedure?
a) To prevent accidental contact with live electrical parts
b) To increase the efficiency of the installation process
c) To reduce the cost of the installation
d) To decrease the overall safety of the installation

Answer: a) To prevent accidental contact with live electrical parts

Explanation: A lockout/tagout procedure is used to ensure that electrical equipment is properly de-energized and locked out before any maintenance or repair work is done. This helps prevent accidental contact with live electrical parts, which can be dangerous or even deadly.

Which of the following is a key consideration when installing electrical wiring in a hazardous location?
a) The type of conduit used
b) The size of the conductors used
c) The location of the wiring in relation to flammable materials
d) The type of insulation used on the conductors

Answer: c) The location of the wiring in relation to flammable materials

Explanation: When installing electrical wiring in a hazardous location, it is important to consider the location of the wiring in relation to flammable materials, as well as the type of conduit and insulation used. Specialized materials and installation techniques may be required to ensure safety and compliance with codes and regulations.

What is the purpose of a megohmmeter in an electrical installation?
a) To measure the resistance of a conductor
b) To test the continuity of a circuit
c) To measure the voltage of a circuit
d) To test the insulation resistance of a conductor

Answer: d) To test the insulation resistance of a conductor

Explanation: A megohmmeter is a specialized instrument used to test the insulation resistance of conductors and other electrical components. This helps ensure that the insulation is working properly and that the component is safe to use in an electrical system.

Which of the following is a common maintenance task for electrical equipment?
a) Replacing the equipment with a new one
b) Cleaning the equipment
c) Disposing of the equipment
d) Ignoring the equipment until it fails

Answer: b) Cleaning the equipment

Explanation: Cleaning electrical equipment is a common maintenance task to ensure it is free of dust and debris that can affect its performance and lifespan.

Which of the following can be a potential hazard during electrical maintenance?
a) Wearing personal protective equipment
b) Working alone
c) Using the correct tools and equipment
d) Following lockout/tagout procedures

Answer: b) Working alone

Explanation: Working alone during electrical maintenance can be a potential hazard as there may not be anyone around to assist in case of an emergency.

Which of the following is a recommended frequency for electrical maintenance?
a) Every 10 years
b) Every 5 years
c) Every 2 years
d) Every year

Answer: d) Every year

Explanation: Electrical maintenance should be done at least once a year to ensure equipment is functioning properly and safely.

Which of the following is a good practice during electrical maintenance?
a) Keeping all equipment plugged in during maintenance
b) Wearing jewelry and loose clothing
c) Using non-insulated tools
d) Disconnecting the equipment from the power source

Answer: d) Disconnecting the equipment from the power source

Explanation: It is important to disconnect the equipment from the power source during maintenance to prevent electrical shock and other hazards.

Which of the following can be a potential hazard during electrical maintenance?
a) Properly labeling equipment
b) Using lockout/tagout procedures
c) Wearing non-conductive shoes
d) Using damaged or worn out equipment

Answer: d) Using damaged or worn out equipment

Explanation: Using damaged or worn out equipment during maintenance can be a potential hazard as it may not function properly and can cause injuries or damage.

What should be the first step when repairing an electrical circuit?

a. Test for voltage

b. Disconnect the circuit

c. Remove all the wires

d. Replace the circuit breaker

Answer: a. Test for voltage

Explanation: Before repairing an electrical circuit, it's important to test for voltage to ensure that the circuit is not live and can be safely repaired. Failure to do so can result in electrocution or damage to equipment.

What is the purpose of a continuity test in electrical repair?

a. To test for voltage

b. To test for resistance

c. To test for current

d. To test for a complete circuit

Answer: d. To test for a complete circuit

Explanation: A continuity test checks for the presence of a complete circuit between two points. This test is useful in identifying broken wires or bad connections.

What tool is commonly used for cutting and stripping wires during electrical repair?

a. Pliers

b. Screwdriver

c. Wire stripper

d. Hammer

Answer: c. Wire stripper

Explanation: Wire strippers are used for cutting and stripping wires during electrical repair. This tool helps to ensure that the wire is not damaged during the process.

What is the purpose of a multimeter in electrical repair?

a. To measure voltage, current, and resistance
b. To test for voltage only
c. To test for current only
d. To test for resistance only

Answer: a. To measure voltage, current, and resistance

Explanation: A multimeter is a versatile tool used in electrical repair that can measure voltage, current, and resistance. This tool is useful in identifying electrical problems and diagnosing issues in circuits and equipment.

What is the purpose of a fuse in electrical repair?
a. To protect equipment from overvoltage
b. To protect equipment from undervoltage
c. To protect equipment from high current
d. To protect equipment from low current

Answer: c. To protect equipment from high current

Explanation: Fuses are used to protect equipment from high currents that could damage or destroy the equipment. When a fuse detects a high current, it will break the circuit and prevent further damage to the equipment.

Which of the following can cause a circuit breaker to trip repeatedly?
a) Overloaded circuit
b) Short circuit
c) Ground fault
d) All of the above

Answer: d

Explanation: An overloaded circuit, short circuit, or ground fault can all cause a circuit breaker to trip repeatedly. It's important to identify and fix the root cause of the problem to prevent further damage or safety hazards.

What is the minimum distance required between buried electrical cables and gas pipes according to NEC?
a) 6 inches
b) 12 inches
c) 18 inches
d) 24 inches

Answer: b
Explanation: According to NEC, the minimum distance required between buried electrical cables and gas pipes is 12 inches. This helps prevent electrical current from flowing through the gas pipe and causing damage or injury.

What is a common cause of a low voltage condition in an electrical system?
a) Overloaded circuit
b) High resistance connection
c) Short circuit
d) Ground fault

Answer: b
Explanation: A high resistance connection is a common cause of a low voltage condition in an electrical system. This can be caused by loose connections, damaged or corroded wires, or other factors.

Which of the following is an important step in troubleshooting an electrical system?
a) Disassembling the equipment
b) Testing every component
c) Replacing all fuses
d) Observing and documenting the problem

Answer: d

Explanation: Observing and documenting the problem is an important step in troubleshooting an electrical system. This helps identify patterns or underlying causes of the problem, which can then be addressed more effectively.

What is the purpose of a megohmmeter in troubleshooting an electrical system?
a) To measure voltage levels
b) To measure current levels
c) To measure resistance levels
d) To measure insulation resistance

Answer: d
Explanation: A megohmmeter is used to measure insulation resistance in an electrical system. This helps identify potential problems with wiring or insulation, which can then be addressed before they cause further damage or safety hazards.

What does a dashed line on an electrical blueprint indicate?
a) A ground wire
b) A circuit breaker
c) A switch
d) A conduit

Answer: a) A ground wire
Explanation: Dashed lines on an electrical blueprint are commonly used to indicate ground wires.

What does an upward pointing arrow with the letter "V" indicate on an electrical blueprint?
a) A voltage source
b) A current source
c) A transformer
d) A capacitor

Answer: a) A voltage source
Explanation: An upward pointing arrow with the letter "V" is a common symbol used to indicate a voltage source on an electrical blueprint.

What does a rectangle with the letter "M" inside it indicate on an electrical schematic?
a) A motor
b) A switch
c) A fuse
d) A transformer

Answer: a) A motor
Explanation: A rectangle with the letter "M" inside it is a common symbol used to indicate a motor on an electrical schematic.

What does a circle with the letter "L" inside it indicate on an electrical blueprint?
a) A load
b) A light
c) A fuse
d) A switch

Answer: a) A load
Explanation: A circle with the letter "L" inside it is a common symbol used to indicate a load on an electrical blueprint.

Which of the following is a proper technique for installing electrical wiring in a building?
a) Use unapproved extension cords as permanent wiring
b) Use undersized wiring to save money
c) Run wiring through walls and floors without proper protection
d) Use approved wiring and conduit to protect wiring from damage

Answer: d) Use approved wiring and conduit to protect wiring from damage

Explanation: The National Electrical Code (NEC) requires that all wiring be protected from damage and installed in approved conduit. Using unapproved extension cords as permanent wiring, undersized wiring, and running wiring without proper protection can be hazardous and a violation of the code.

What is the purpose of a ground fault circuit interrupter (GFCI)?
a) To prevent electrical fires
b) To protect sensitive electronic equipment from power surges
c) To protect users from electrical shock
d) To provide a stable source of electricity to a building

Answer: c) To protect users from electrical shock

Explanation: A GFCI is a device that monitors the flow of current in a circuit and shuts off power if it detects an imbalance in current flow. This imbalance can indicate that current is flowing through a person instead of the intended path, which can cause electrical shock. GFCIs are required by the NEC in certain locations, such as bathrooms, kitchens, and outdoor areas, to help prevent electrical shock.

Which of the following is an example of a proper installation of a light fixture?
a) Wiring the fixture without turning off the power
b) Using an undersized wire to connect the fixture to the electrical system
c) Attaching the fixture to an unsupported ceiling box
d) Securing the fixture to a properly installed ceiling box

Answer: d) Securing the fixture to a properly installed ceiling box

Explanation: The NEC requires that all light fixtures be securely attached to an approved ceiling box to prevent the fixture from falling. Wiring a fixture without turning off the power, using an undersized wire, and attaching a fixture to an unsupported ceiling box can all be hazardous and are not in compliance with the code.

What is the maximum number of conductors allowed in a single conduit?
a) 2
b) 3
c) 4
d) 6

Answer: d) 6

Explanation: The NEC specifies that the maximum number of conductors allowed in a single conduit depends on the size of the conduit and the size of the conductors. In general, a conduit can contain up to 6 conductors, but this number may be lower for larger conductors or smaller conduit sizes.

What is the purpose of a disconnect switch?
a) To protect users from electrical shock
b) To protect electrical equipment from damage
c) To provide a safe way to turn off power to a circuit or piece of equipment
d) To regulate the voltage of an electrical system

Answer: c) To provide a safe way to turn off power to a circuit or piece of equipment

Explanation: A disconnect switch is a device that allows users to safely turn off power to a circuit or piece of equipment without touching any live wires. Disconnect switches are required by the NEC in certain applications, such as for air conditioning and heating equipment, to provide a safe way to turn off power during maintenance or repairs.

What is the first step in troubleshooting an electrical circuit?
a) Check the voltage at the load
b) Inspect the circuit breakers or fuses
c) Check the continuity of the circuit
d) Verify that the equipment is turned on

Answer: d) Verify that the equipment is turned on
Explanation: Before starting any troubleshooting, it's important to verify that the equipment is turned on and in working order. This can save time by eliminating unnecessary tests or inspections.

Which of the following tools is commonly used for troubleshooting electrical circuits?
a) Voltage tester
b) Screwdriver
c) Tape measure
d) Pliers

Answer: a) Voltage tester
Explanation: A voltage tester is a commonly used tool for troubleshooting electrical circuits. It's used to measure the voltage at different points in the circuit to identify where the problem may be.

What is the purpose of a ground fault circuit interrupter (GFCI)?
a) To prevent electrical fires
b) To protect against electric shock
c) To reduce electrical noise
d) To regulate the voltage in a circuit

Answer: b) To protect against electric shock
Explanation: A GFCI is designed to protect against electric shock by monitoring the flow of current in a circuit. If the current flowing out of the circuit is not equal to the current flowing back in, the GFCI will trip and shut off power to the circuit.

When troubleshooting an electrical circuit, what is the next step after verifying that the equipment is turned on?
a) Check the voltage at the load
b) Inspect the circuit breakers or fuses
c) Check the continuity of the circuit
d) Test for shorts or grounds

Answer: b) Inspect the circuit breakers or fuses
Explanation: After verifying that the equipment is turned on and in working order, the next step in troubleshooting an electrical circuit is to inspect the circuit breakers or fuses to ensure that they are not tripped or blown.

What is a common cause of electrical faults in a residential wiring system?

a) Overloading the circuit with too many appliances

b) Use of low-quality wiring materials

c) Incorrect wiring connections

d) Failure to follow building codes

Answer: c) Incorrect wiring connections

Explanation: Incorrect wiring connections are a common cause of electrical faults in residential wiring systems. It's important to follow proper wiring procedures and to use quality materials to ensure safe and reliable electrical installations.

Which of the following codes is used to regulate electrical installations in the United States?

a) IEC

b) NEC

c) ANSI

d) IEEE

Answer: b) NEC

Explanation: The National Electrical Code (NEC) is the most widely used standard in the United States to regulate electrical installations.

Which of the following codes is used for the design of high-performance green buildings?

a) NEC

b) ASHRAE 90.1

c) NFPA 70E

d) OSHA

Answer: b) ASHRAE 90.1

Explanation: The American Society of Heating, Refrigerating and Air-Conditioning Engineers (ASHRAE) 90.1 standard provides guidelines for the design of high-performance green buildings.

Which of the following codes is used to ensure fire safety in buildings?

a) NEC

b) ASHRAE 90.1
c) NFPA 70E
d) NFPA 101

Answer: d) NFPA 101
Explanation: The National Fire Protection Association (NFPA) 101 standard provides guidelines for ensuring fire safety in buildings.

Which of the following codes provides guidelines for electrical safety in the workplace?
a) NEC
b) ASHRAE 90.1
c) NFPA 70E
d) OSHA

Answer: c) NFPA 70E

Explanation: The NFPA 70E standard provides guidelines for electrical safety in the workplace.

Which of the following codes is used to ensure safety of electrical equipment in hazardous locations?
a) NEC
b) ASHRAE 90.1
c) NFPA 70E
d) NFPA 70

Answer: d) NFPA 70

Explanation: The NFPA 70 standard, also known as the National Electrical Code (NEC), provides guidelines for the safe installation of electrical equipment in hazardous locations.

Which of the following is an example of special occupancy equipment?
a) HVAC system
b) Lighting fixtures
c) Security cameras

d) None of the above

Answer: a) HVAC system

Explanation: Special occupancy equipment refers to equipment that is designed for specific types of occupancy or specific conditions, such as hazardous locations, medical facilities, or industrial processes. HVAC (heating, ventilation, and air conditioning) systems are an example of special occupancy equipment, as they are designed for a specific type of occupancy and require special installation and maintenance.

In which of the following occupancies would you typically find Class I hazardous locations?
a) Office buildings
b) Residential buildings
c) Hospitals
d) Chemical processing plants

Answer: d) Chemical processing plants

Explanation: Class I hazardous locations are areas where flammable gases or vapors are present in the air in sufficient quantities to ignite or explode. These locations are commonly found in chemical processing plants, refineries, and other industrial facilities where hazardous materials are stored, processed, or handled.

Which of the following is a requirement for electrical equipment used in damp locations?
a) GFCI protection
b) Insulation resistance testing
c) Corrosion-resistant materials
d) Explosion-proof enclosures

Answer: c) Corrosion-resistant materials

Explanation: Electrical equipment used in damp locations must be designed and constructed to prevent moisture from entering the equipment and causing damage. This includes the use of corrosion-resistant materials that can withstand exposure to moisture and other environmental factors.

Which of the following is a requirement for electrical equipment used in patient care areas?
a) Ground-fault circuit interrupter protection
b) Explosion-proof enclosures
c) Class I, Division 1 rating
d) None of the above

Answer: a) Ground-fault circuit interrupter protection

Explanation: Electrical equipment used in patient care areas must be designed and installed to minimize the risk of electrical shock to patients and staff. This includes the use of ground-fault circuit interrupter (GFCI) protection to detect and interrupt electrical current if a fault is detected.

Which of the following is an example of a special condition that may affect the installation of electrical equipment?
a) High ambient temperatures
b) High humidity levels
c) Dust and debris
d) All of the above

Answer: d) All of the above

Explanation: Special conditions that may affect the installation of electrical equipment include high ambient temperatures, high humidity levels, and the presence of dust and debris. These conditions can affect the performance and reliability of electrical equipment and must be taken into account during installation and maintenance.

Which of the following formulas is used to calculate power in an electrical circuit?
a) $P = VI$
b) $P = V^2/R$

c) P = I^2R

d) P = Q/T

Answer: b)

Explanation: The formula to calculate power in an electrical circuit is P = V^2/R, where V is the voltage and R is the resistance.

What is the formula to calculate the total resistance in a series circuit?

a) R = R1 + R2 + R3

b) R = 1/R1 + 1/R2 + 1/R3

c) R = R1R2R3

d) R = V/I

Answer: a)

Explanation: In a series circuit, the total resistance is calculated by adding the individual resistances together. The formula is R = R1 + R2 + R3, where R1, R2, and R3 are the individual resistances in the circuit.

Which formula is used to calculate the current in a circuit?

a) I = V/R

b) I = PR

c) I = P/V

d) I = V/Q

Answer: a)

Explanation: The formula to calculate the current in a circuit is I = V/R, where V is the voltage and R is the resistance.

What is the formula to calculate the power factor in an AC circuit?

a) PF = V/I

b) PF = P/VI

c) PF = P/VA

d) PF = VA/P

Answer: c)

Explanation: The power factor in an AC circuit is calculated using the formula PF = P/VA, where P is the real power and VA is the apparent power.

What is the formula to calculate the capacitance of a capacitor?
a) C = Q/V
b) C = V/Q
c) C = Q^2/V
d) C = V^2/Q

Answer: a)

Explanation: The capacitance of a capacitor is calculated using the formula C = Q/V, where Q is the charge on the capacitor and V is the voltage across it.

What is the minimum safe working distance from a live 480-volt electrical panel?
a) 1 foot
b) 3 feet
c) 6 feet
d) 10 feet

Answer: c) 6 feet

Explanation: According to the OSHA standard 1910.333(a), the minimum safe working distance from a live 480-volt electrical panel is 6 feet.

What is the purpose of lockout/tagout procedures?
a) To prevent unauthorized access to electrical equipment
b) To prevent the use of faulty electrical equipment
c) To prevent electrical shock and other injuries during maintenance and repair
d) To prevent electrical fires in equipment

Answer: c) To prevent electrical shock and other injuries during maintenance and repair

Explanation: Lockout/tagout procedures are used to prevent the release of hazardous energy and to protect workers from electrical shock and other injuries during maintenance and repair of electrical equipment.

Which of the following is an example of a Class C fire?
a) Electrical fire
b) Oil fire
c) Paper fire
d) Wood fire

Answer: a) Electrical fire
Explanation: Class C fires involve energized electrical equipment, and they require special extinguishing agents that are non-conductive and do not leave a residue.

What is the purpose of GFCI protection?
a) To protect against overcurrent conditions
b) To protect against short circuits
c) To protect against ground faults
d) To protect against lightning strikes

Answer: c) To protect against ground faults
Explanation: Ground fault circuit interrupter (GFCI) protection is designed to protect against ground faults, which occur when electricity leaks from a circuit and flows through a person or an object to reach the ground.

What is the maximum voltage rating for a GFCI receptacle?
a) 120 volts
b) 240 volts
c) 277 volts
d) 480 volts

Answer: a) 120 volts

Explanation: The maximum voltage rating for a GFCI receptacle is 120 volts, as specified in the NEC Article 210.8.

Which of the following renewable energy technologies converts sunlight directly into electricity?
a) Wind turbines
b) Hydroelectric power
c) Geothermal energy
d) Solar photovoltaic systems

Answer: d) Solar photovoltaic systems

Explanation: Solar photovoltaic systems use solar panels to convert sunlight directly into electricity, while wind turbines generate electricity from wind energy, hydroelectric power is generated from flowing water, and geothermal energy is heat from the Earth's core.

Which of the following is a type of energy storage technology that can be used to store excess energy generated by renewable sources?
a) Flywheels
b) Capacitors
c) Batteries
d) Superconductors

Answer: c) Batteries

Explanation: Batteries are commonly used to store excess energy generated by renewable sources such as solar or wind power. Flywheels, capacitors, and superconductors are other types of energy storage technologies, but they are not as commonly used for this purpose.

Which of the following renewable energy technologies is a form of bioenergy?
a) Wind power
b) Solar power
c) Hydroelectric power
d) Biomass energy

Answer: d) Biomass energy

Explanation: Biomass energy involves using organic materials such as wood, crops, or waste to generate heat or electricity. Wind power, solar power, and hydroelectric power are all forms of energy that do not involve organic materials.

Which of the following renewable energy technologies is typically used in off-grid applications where grid electricity is not available?
a) Geothermal energy
b) Wind power
c) Solar power
d) Hydroelectric power

Answer: c) Solar power

Explanation: Solar power is often used in off-grid applications such as remote cabins, RVs, or marine vessels, where grid electricity is not available. Geothermal energy, wind power, and hydroelectric power are typically used in grid-connected applications.

Which of the following is a common way to increase the efficiency of a solar photovoltaic system?
a) Using larger solar panels
b) Adding more batteries
c) Increasing the tilt angle of the panels
d) Using a lower-capacity inverter

Answer: c) Increasing the tilt angle of the panels

Explanation: Increasing the tilt angle of the solar panels can improve the efficiency of a solar photovoltaic system by maximizing the amount of sunlight that hits the panels. Using larger panels or more batteries can increase the system's capacity but does not necessarily increase efficiency. Using a lower-capacity inverter may actually decrease efficiency.

Which NEC article covers general requirements for overcurrent protection?

a) Article 240
b) Article 250
c) Article 430
d) Article 480

Answer: a) Article 240

Explanation: NEC Article 240 covers general requirements for overcurrent protection. It includes requirements for the selection and installation of overcurrent protective devices, conductor sizing for overcurrent protection, and coordination of overcurrent protective devices.

What is the formula for calculating the current-carrying capacity of a conductor?
a) $I = P/V$
b) $I = (kVA \times 1000) / V$
c) $I = (2 \times k \times L \times A) / (1000 \times C)$
d) $I = (A \times 1000) / k$

Answer: d) $I = (A \times 1000) / k$
Explanation: The formula for calculating the current-carrying capacity of a conductor is $I = (A \times 1000) / k$, where I is the current in amperes, A is the cross-sectional area of the conductor in square millimeters, and k is the specific resistance of the conductor material in ohms per kilometer.

What is the maximum voltage drop allowed by NEC for branch circuit conductors?
a) 1%
b) 2%
c) 3%
d) 5%

Answer: b) 2%
Explanation: According to NEC Article 210, branch circuit conductors shall have an allowable voltage drop not exceeding 2% of the nominal voltage of the circuit.

What is the minimum clearance required by NEC between overhead conductors and the ground or other objects?
a) 8 feet
b) 10 feet
c) 12 feet
d) 14 feet

Answer: b) 10 feet

Explanation: According to NEC Article 225, overhead conductors must have a vertical clearance of not less than 10 feet above the surface of the ground, sidewalks, or platforms.

What is the formula for calculating the size of the equipment grounding conductor required for a circuit?
a) EGC = (IL x 0.2) / Cmin
b) EGC = (IL x 0.8) / Cmin
c) EGC = (IL x 1.2) / Cmin
d) EGC = (IL x 2.0) / Cmin

Answer: a) EGC = (IL x 0.2) / Cmin

Explanation: The formula for calculating the size of the equipment grounding conductor required for a circuit is EGC = (IL x 0.2) / Cmin, where EGC is the size of the equipment grounding conductor in circular mils, IL is the current in amperes, and Cmin is the minimum circular mil area of the equipment grounding conductor as determined by NEC Table 250.122.

Which NEC article covers requirements for services?
a) Article 220
b) Article 230
c) Article 240
d) Article 250

Answer: b) Article 230. This article covers service conductors and equipment, and the requirements for electrical service installations.

Which NEC article covers requirements for branch circuits?
a) Article 220
b) Article 230
c) Article 240
d) Article 210

Answer: d) Article 210. This article covers the general requirements for branch circuits, including conductor sizing, overcurrent protection, and installation methods.

What is the maximum number of current-carrying conductors allowed in a conduit per NEC 310.15(B)(3)(a)?
a) 1
b) 3
c) 9
d) 20

Answer: b) 3. NEC 310.15(B)(3)(a) sets the limit for the number of current-carrying conductors in a conduit or raceway based on their size and the insulation type.

Which NEC article covers requirements for grounding and bonding?
a) Article 220
b) Article 230
c) Article 250
d) Article 300

Answer: c) Article 250. This article covers the requirements for grounding and bonding of electrical systems, equipment, and structures.

What is the minimum height requirement for outdoor service conductors per NEC 230.9?
a) 10 feet
b) 12 feet
c) 15 feet
d) 18 feet

Answer: b) 12 feet. NEC 230.9 requires that outdoor service conductors be installed at a height of not less than 12 feet above finished grade. This is to help ensure that people and vehicles cannot come into contact with the conductors.

What is the minimum size of the equipment grounding conductor required for a 400-ampere, 3-phase, 480-volt feeder?
a. 2/0 AWG
b. 1/0 AWG
c. 4 AWG
d. 8 AWG

Answer: b. 1/0 AWG
Explanation: NEC 250.122 specifies that the equipment grounding conductor for a feeder rated 401 to 600 amperes must not be smaller than 1/0 AWG.

What is the maximum allowable voltage drop for feeders and branch circuits as per NEC?
a. 2%
b. 3%
c. 5%
d. 10%

Answer: b. 3%
Explanation: According to NEC 210.19(A)(1), the maximum allowable voltage drop for feeders and branch circuits is 3% of the nominal voltage of the circuit.

What is the minimum ampacity of the conductors for a 20-ampere, 120-volt circuit?
a. 14 AWG
b. 12 AWG
c. 10 AWG
d. 8 AWG

Answer: a. 14 AWG

Explanation: NEC 310.16 specifies that 14 AWG conductors are rated for 15 amperes, which is sufficient for a 20-ampere circuit.

According to NEC, what is the minimum headroom clearance required above service equipment?

a. 2.5 feet

b. 3 feet

c. 6 feet

d. 8 feet

Answer: c. 6 feet

Explanation: NEC 110.26 specifies that a minimum headroom clearance of 6 feet is required above service equipment.

What is the maximum allowable number of conductors in a single raceway for power-limited fire alarm circuits?

a. 1

b. 2

c. 3

d. 4

Answer: d. 4

Explanation: NEC 760.28 specifies that the maximum allowable number of conductors in a single raceway for power-limited fire alarm circuits is 4.

What is the minimum size conduit required for a circuit with four #10 THHN wires and one #8 THHN ground wire?

a) 1/2 inch

b) 3/4 inch

c) 1 inch

d) 1-1/4 inch

Answer: b) 3/4 inch

Explanation: NEC 310.15(B)(16) requires a minimum of 3/4 inch conduit for a circuit with four #10 THHN wires and one #8 THHN ground wire.

What is the minimum radius of a conduit bend for a 1-inch conduit?
a) 4 inches
b) 6 inches
c) 8 inches
d) 10 inches

Answer: b) 6 inches
Explanation: NEC 352.26 requires a minimum radius of 6 times the conduit diameter for conduit bends.

What is the maximum number of 12 AWG conductors allowed in a 1/2 inch conduit?
a) 6
b) 8
c) 10
d) 12

Answer: b) 8
Explanation: NEC Table 1 in Chapter 9 provides the maximum number of conductors allowed in various sizes of conduit, and it specifies that a maximum of 8 12 AWG conductors are allowed in a 1/2 inch conduit.

What is the minimum cover depth required for a PVC conduit installed under a driveway?
a) 6 inches
b) 12 inches
c) 18 inches
d) 24 inches

Answer: b) 12 inches

Explanation: NEC 300.5 specifies the minimum cover depths required for various types of conduit installations, and it requires a minimum of 12 inches of cover for PVC conduit installed under a driveway.

What is the minimum size conduit required for a circuit with three #6 THHN wires and one #10 THHN ground wire?
a) 1/2 inch
b) 3/4 inch
c) 1 inch
d) 1-1/4 inch

Answer: c) 1 inch
Explanation: NEC 310.15(B)(16) requires a minimum of 1 inch conduit for a circuit with three #6 THHN wires and one #10 THHN ground wire.

What is the maximum allowable voltage drop in a feeder or branch circuit, as per NEC code?
a) 1%
b) 2%
c) 3%
d) 4%

Answer: b) 2%
Explanation: As per NEC code 210.19(A)(1), the maximum allowable voltage drop for a feeder or branch circuit should not exceed 2%.

According to NEC code 310.15(B)(16), what is the ampacity of a 12 AWG copper conductor rated for 90°C when installed in a raceway with three current-carrying conductors?
a) 20A
b) 25A
c) 30A
d) 35A

Answer: a) 20A

Explanation: As per NEC code 310.15(B)(16), a 12 AWG copper conductor rated for 90°C can carry 20A when installed in a raceway with three current-carrying conductors.

As per NEC code 215.2(A)(1), what is the minimum feeder size for a continuous load of 80A?
a) 4 AWG copper
b) 2 AWG copper
c) 1/0 AWG copper
d) 3/0 AWG copper

Answer: d) 3/0 AWG copper
Explanation: As per NEC code 215.2(A)(1), the minimum feeder size for a continuous load of 80A is 3/0 AWG copper.

According to NEC code 310.16, what is the ampacity of a 4/0 AWG aluminum conductor rated for 90°C and installed in free air?
a) 185A
b) 200A
c) 225A
d) 250A

Answer: c) 225A
Explanation: As per NEC code 310.16, a 4/0 AWG aluminum conductor rated for 90°C and installed in free air can carry 225A.

What is the minimum size equipment grounding conductor required for a 200A service entrance with parallel 500 kcmil aluminum phase conductors, as per NEC code 250.122?
a) 1/0 AWG copper
b) 2/0 AWG copper
c) 3/0 AWG copper
d) 4/0 AWG copper

Answer: b) 2/0 AWG copper
Explanation: As per NEC code 250.122, the minimum size equipment grounding conductor required for a 200A service entrance with parallel 500 kcmil aluminum phase conductors is 2/0 AWG copper.

What is the NEC-recommended minimum load for a dwelling unit?
a) 100 amps
b) 125 amps
c) 150 amps
d) 200 amps

Answer: b) 125 amps
Explanation: According to NEC Article 220, the minimum calculated load for a dwelling unit is 125 amps.

What is the NEC requirement for the calculation of commercial loads?
a) Detailed calculations are required for all commercial projects.
b) Simplified calculation methods are allowed for smaller commercial projects.
c) Only licensed engineers are allowed to perform commercial load calculations.
d) Load calculations are not required for commercial projects.

Answer: b) Simplified calculation methods are allowed for smaller commercial projects.
Explanation: NEC Article 220 allows for simplified calculation methods for commercial projects that meet certain size and use criteria.

What is the NEC requirement for demand factors in residential load calculations?
a) Demand factors must be applied to all loads.
b) Demand factors are not required for residential load calculations.
c) Demand factors only apply to heating and air conditioning loads.
d) Demand factors only apply to lighting and appliance loads.

Answer: d) Demand factors only apply to lighting and appliance loads.
Explanation: According to NEC Article 220, demand factors are only applied to lighting and appliance loads in residential load calculations.

What is the NEC-recommended minimum voltage for residential services?
a) 110 volts
b) 120 volts

c) 240 volts

d) 480 volts

Answer: b) 120 volts

Explanation: The NEC recommends a minimum voltage of 120 volts for residential services.

What is the NEC requirement for the sizing of service-entrance conductors in a commercial building?

a) The conductors must be sized based on the calculated load of the building.

b) The conductors must be sized based on the highest calculated load in the building.

c) The conductors must be sized based on the maximum overcurrent protection device.

d) The conductors must be sized based on the voltage drop of the circuit.

Answer: b) The conductors must be sized based on the highest calculated load in the building.

Explanation: According to NEC Article 310, service-entrance conductors for a commercial building must be sized based on the highest calculated load in the building.

Printed in the USA
CPSIA information can be obtained
at www.ICGtesting.com
LVHW050900300923
759516LV00017B/597

9 781088 203132